Totalitarian
Science and
Technology

THE CONTROL OF NATURE
Series Editors:
Margaret C. Jacob, Rosemary Stevens,
and Spencer R. Weart

PUBLISHED

SCIENTISTS AND THE DEVELOPMENT OF NUCLEAR WEAPONS
From Fission to the Limited Test Ban Treaty, 1939–1963
Lawrence Badash

EINSTEIN AND OUR WORLD
David Cassidy

NEWTON AND THE CULTURE OF NEWTONIANISM
Betty Jo Teeter Dobbs and Margaret C. Jacob

CONTROLLING HUMAN HEREDITY
1865 to the Present
Diane B. Paul

TOTALITARIAN SCIENCE AND TECHNOLOGY
Paul R. Josephson

FORTHCOMING

SCIENCE AND TECHNOLOGY IN THE INDUSTRIAL REVOLUTION
Eric Brose

GENDER AND SCIENCE
Paula Findlen and Michael Dietrich

THE SCIENTIFIC REVOLUTION FROM COPERNICUS TO NEWTON
James Jacob

Totalitarian
Science and
Technology

Paul R. Josephson

HUMANITIES PRESS
NEW JERSEY

First published in 1996 by Humanities Press International, Inc.
165 First Avenue, Atlantic Highlands, New Jersey 07716

©1996 by Paul R. Josephson

Library of Congress Cataloging–in–Publication Data

Josephson, Paul R.
    Totalitarian science and technology / Paul R. Josephson.
       p. cm. — (Control of nature)
    Includes bibliographical references and index.
    ISBN 0–391–03979–2 (cloth). — ISBN 0–391–03980–6 (paper)
    1. Science—Social aspects—Germany—History—20th century.
  2. Science—Political aspects—Germany—History—20th century.
  3. Technology—Social aspects—Germany—History—20th century.
  4. Technology—Political aspects—Germany—History—20th century.
  5. Totalitarism.   6. Hitler, Adolph, 1889–1945.   7. Stalin, Joseph,
1879–1953.   I. Title.   II. Series.
Q175.52.G3J67   1996
306.4′5′094309043—dc20   96–11772
                                     CIP

Printed in the United States of America

10   9   8   7   6   5   4   3   2   1

Contents

v

Series Editors' Preface

THIS SERIES OF historical studies aims to enrich understanding of the role that science and technology have played in the history of Western civilization and culture, and through that in the emerging modern world civilization. Each author has written with students and general readers, not specialists, in mind. And the volumes have been written by scholars distinguished in the particular field. In this book Paul Josephson—well-known as one of the very few historians who has attained access to archives and interviewed scientists in the Russian physics community—draws on his expertise to address key questions about the relations among science, technology and political systems under the most extreme conditions.

The aim of this book on totalitarian science and technology is not just to lay out some basic historical information, which could only be a sample of the many complex developments that scholars are currently exploring. Still more this volume intends to show the chief questions and debates that inform current historical scholarship.

The current debates as presented here emphasize the "Control of Nature." While not excluding a discussion of how knowledge itself develops, how it is constructed through the interplay of research into nature with the values and beliefs of the researcher, this volume—like all the others in the series—looks primarily at how science and technology interact with economic, social, linguistic and intellectual life, in ways that transform the relationship between human beings and nature. In every volume we are asking the student to think about how the modern world came to be invented, a world where the call for progress and the need to respect humanity and nature produce a tension, on the one hand liberating, on the other threatening to overwhelm human resources and ingenuity. The scientists whom you will meet here could not in every case have foreseen the kind of power that modern science and technology now offer. But they were also dreamers and doers—as well as shrewd promoters—who changed forever the way people view the natural world.

MARGARET C. JACOB
ROSEMARY STEVENS
SPENCER WEART

vii

Acknowledgments

I would like to acknowledge the help of several individuals in seeing this book come to fruition. Peg Jacob and Spencer Weart asked me several years ago to consider writing a textbook for this series. They encouraged me at every step of the way once I set forth the basic idea for this text. I welcome Spencer's thorough editorial work. Gene Rinchik helped with chapter 2. Paul Forman offered important comments on chapter 3. Tom Hughes provided detailed comments on chapter 4. Tom Gleason shared his manuscript on totalitarianism with me and encouraged clarification of my ideas on totalitarianism. Blair Ruble shared with me his work on urban planning in Moscow. Katie Lippa read chapter 2 carefully. Miriam Conant made suggestions on an early version of chapter 1. My son, Isaac, rarely insisted that I cross-country ski with him before 6:30 A.M., so that I could have some time to write in the mornings. I dedicate this book to my good friend and colleague Viktor Frenkel, with whom I have had the good fortune to share my work on the fate of scientists in the former Soviet Union for nearly two decades.

The reader need not be a practicing scientist or have extensive laboratory experience to make good use of this book. Some background will be assumed; in places I will provide discussion of various scientific disciplines necessary to full comprehension of the arguments. But inasmuch as I argue that science is socially and politically constructed, the task at hand for the reader is to understand what impact state policies in totalitarian regimes had on science that distinguishes it from the same discipline in another culture and polity.

Totalitarian Science

ADOLF HITLER AND Joseph Stalin enlisted scientists and engineers in their efforts to build strong states. They desired industrial power and military might. To these ends they underwrote expensive research and development (R and D) in scientific institutes. Hitler loved "superweapons." Stalin closely followed the Soviet atomic bomb project. Some of the scientists who toiled for Stalin were Nikolai Vavilov, an internationally renowned biologist who died in a labor camp while his brother became president of the Academy of Sciences, and Andrei Sakharov, father of the Soviet hydrogen bomb and later a political dissident. Hitler's stable of specialists included Werner Heisenberg, one of the founders of quantum mechanics.

When one thinks about science and technology in totalitarian regimes like Nazi Germany, horrible images of doctors undertaking concentration camp experiments on unwilling prisoners come to mind. Another image of the kind of science possible under a dictatorship is that of Lysenkoism in the Soviet Union, named after Trofim Lysenko, who controlled biological research from 1935 until 1965 and used his power to require the rejection of modern genetics. We recall how Andrei Sakharov and the leading Chinese theoretical physicist, Fang Lizhi, spent years in exile or under house arrest for openly criticizing their governments' human rights records (Fang 1990).

But are these cases representative of what constitutes science in totalitarian regimes, or are they anomalies? Several leading historians and sociologists of science maintain that science operates according to democratic principles. They claim that these principles prevent any members of the scientific community from establishing their views about the phenomena at hand as sacrosanct and ensure that scientific discovery is an adversarial and cumulative process that brings us ever closer to the "truth." In an essay published in 1942 directed in part against the anti-intellectualism of totalitarian regimes, Robert Merton

1

argued that the ethos of science, its universalism, communal character, disinterestedness, and organized skepticism, militated against particular dogmas of church, economy, and state. Twenty years later Michael Polanyi acknowledged that orthodoxy existed in science but called it a "dynamic orthodoxy." He argued that "the authority in scientific opinion remains essentially mutual; it is established *between* scientists, not above them," and that a "republic of science" existed to mediate disputes. Indeed, Polanyi, Karl Popper, and other philosophers of science wrote about this dynamic orthodoxy in part as an attack on pseudo-science, Lysenkoism, and so on (Merton 1973; Polanyi 1962; Popper 1945).

This view of science is hard to reconcile with the fact that in a number of totalitarian systems the science enterprise is dynamic. In the most prominent case, the Soviet Union, not only was society as a whole subjected to arbitrary one-party rule, but scientific institutions came to be dominated by scientific administrators whose administration might be characterized as "stagnant orthodoxy." Yet the USSR orbited the first artificial satellite (Sputnik), developed the Tokamak fusion reactor, and in a number of other fields supported scientists who were recognized as world leaders. In the last days of the Soviet regime, in comments before scientists who had gathered to consider the admission of new members and the transformation of the Academy of Sciences into a less elite institution, the academy's last president, Gurii Marchuk, attacked the notion that science is anything like a democracy, since "truth" is not decided by majority vote (Marchuk 1991).

This book aims at a balanced view of science in totalitarian regimes, going beyond mere attacks on their "pseudo-science." All governments have an impact on science and science policy. Since the rise of Western science in the years 1500 to 1800, the state has played an increasingly important role in the conduct of science (Dobbs and Jacob 1995). In the seventeenth century, rising nation-states underwrote honorific and research-oriented scientific societies and universities in England, France, Denmark, Germany, and Russia. By the late eighteenth century, they funded armaments, mining, and metallurgy endeavors. By 1900, governments recognized the importance of science for health, medicine, agriculture, and national defense. They set up land-grant colleges, geological surveys, meteorological services, and national standards facilities. In World War II the relationship between scientists and the state changed forever. Vastly increased government subsidies enabled scientists to develop radar and atomic weapons in national laboratories that after the war became a fixture throughout the world (Josephson 1991; Taubes 1986; Macrakris 1994).

In regimes like the United States, which may be characterized as liberal pluralist regimes, the impact of the state on science is none too subtle. When the U.S. federal government spent billions of dollars on the space race in the 1960s through NASA, or on Star Wars (the Strategic Defense Initiative) in the 1980s, it created demand for tens of thousands of scientists to be trained in such areas as aeronautical engineering, solid-state physics, and computer science and technology; many of them lost their jobs when the projects were cut back. It is the government that decides whether to permit fetal tissue research on aborted or stillborn fetuses in search of cures for various debilitating diseases, which some argue is no more moral than research on humans conducted by the Nazis. Others point to the case of J. Robert Oppenheimer, head of the American atomic bomb project, who was stripped of his security clearance for his vacillation about the development of the hydrogen bomb, as morally equivalent to Andrei Sakharov's treatment at the hands of Soviet leaders. And others criticize the effort of the Reagan administration to secure the right to prior censorship over government-funded research for national security reasons as similar to the behavior of Chinese scientific administrators toward Chinese scientists.

Based on a comparison of science under Hitler and Stalin, however, I will argue that totalitarian regimes have a unique impact on the careers and research interests of scientists and engineers. What is a totalitarian regime?[1] First, there is a monopoly on power usually manifest in one-party rule. A leader or tiny clique presides at the top of the party, with unquestioned and arbitrary personal power. Members of the ruling elite share a fiery commitment to transform society. One of the tools they use is a monistic belief system that encourages the individual to identify with state goals. This belief system, which includes mythical notions of right and wrong, justice and retribution, nationalism, fatherland and/or motherland, and love for the leader, is disseminated through centrally controlled media. The system appeals to instinct as opposed to reason, although claiming the latter. The state employs secret police who use terror, coercion, and violence to reach its aims. It alleges the presence of internal and external enemies to mobilize the masses (Gleason 1995). The future is a fundamental category in the name of which a murderous logic prevails: for example, the thousand-year Reich, or classless society and withering away of the state.

To confuse matters somewhat, totalitarian regimes have allowed some flexibility—individual initiative—in the economic sphere, but less in the political sphere. Nazi Germany had a market-capitalist sector and a state-controlled sector. Engineering in Germany of items not of interest to the state (cameras, say, but not Volkswagens) proceeded largely as in the United States. The People's Republic of China now permits small-scale capitalism to prosper. The USSR had state ownership and control of the economy after brief experimentation with market mechanisms in the mid-1920s to encourage economic recovery from World War I and the Russian Revolution. Soviet leaders then imposed central planning with control over prices and over labor and capital inputs, while permitting initiative among some plant managers and even less private enterprise, notably small family plots of land. However, when individuals tried to organize to oppose state policies or to establish independent political parties, the totalitarian state in all cases acted quickly to destroy any potential for opposition.

These kinds of political controls and ideological constraints notwithstanding, science in totalitarian regimes has generally followed international research paradigms in terms of both focus and the methodologies that scientists employ. Pick up the scientific journals published in Nazi Germany or the Soviet Union, and you will find cutting-edge research in many fields. However, based on an approach in the history and philosophy of science that is called "the social study of science," I will discuss how economic desiderata, political exegencies, and ideological considerations shape the face of totalitarian science, its institutions, and the persons who carry out scientific research.

In chapter 2, which compares the biological sciences in Nazi Germany and the Soviet Union, I argue that a transformationist vision was central to science in these regimes, distinguishing their biological research from that done under other political systems. In the Soviet Union, biology would be used to revolutionize agriculture, to create a "New Soviet Man," even to tame nature within one generation. In Nazi Germany, biology would secure Germany's agricultural self-sufficiency. More important, applied through various race laws, Nazi biology would create a nation of pure Aryan gods whose Reich ruled the world. Granted, Soviet and National Socialist biology held widely different views on, for instance, the role of genetics in determining human qualities. But in both states science served utilitarian ends of transforming social, political, and cultural institutions, as well as nature itself, in short order.

In chapter 3, an analysis of the reception of relativity theory and quantum mechanics (often referred to as "the new physics") in Nazi Germany and the Soviet Union, I argue that the ideologization of science resulted in extrascientific forces coming into play to determine what was "good" science and what was "bad" science. In Nazi Germany and the Soviet Union, the ideologization of science led to the ostracism of physicists who embraced the new physics. In the former, the basis of criticism was racial; in the latter, it was Marxian and class-based. In both cases, individuals who tried to practice science openly and honestly lost their jobs, were arrested, and in some cases were killed for their scientific beliefs.

The ideologization of science succeeds because of an activist state. Science in totalitarian regimes often moves ahead because of the intercession of the state and of extrascientific organizations and individuals that represent it, rather than relying on traditional measures of scientific excellence such as publication in refereed journals, peer review, grant applications, scientific citations, or membership in national and international scientific organizations. In normal practice, scientific disputes are aired openly, although egos may be bruised in the process. Competition between schools of research to determine the validity of a solution to a given scientific problem inspires the confidence of scientists everywhere that they are establishing "facts" independent of political or personal issues. Granted, in pluralist systems such as the United States, scientists go outside of these normal channels to air disputes in the political arena. Controversies over fluoridation of water, what constitutes a wetlands, whether Star Wars antimissile technologies will work, the extent of the greenhouse effect, and so on demonstrate that political, ideological, and economic forces shape scientific debates. But in totalitarian regimes, there are taboo subjects. Researchers who venture into those areas risk job security and personal freedom. Individual scientists, ideologues, and administrators gain the power to define "good" science in a way that proscribes academic freedom.

In chapter 4, I turn to analysis of the nature of technology in totalitarian regimes. Like biology and physics, the development of technology would appear at first glance to reflect objective engineering calculations about the most efficient means to achieve some end. All jets, rockets, automobiles, hydropower stations, and skyscrapers use similar materials and construction techniques and resemble each other physically. Yet there are two features that distinguish large-scale technological systems in totalitarian regimes from those in other systems. The first is that the state is the prime mover in technological development. In order to achieve the goals of economic self-sufficiency and military might, the state harnesses the efforts of engineers and scientists to its programs. It is the main force in shaping what areas merit study. But in exchange for funding, experts are held accountable to produce results, often as specified in national planning documents. Failure to meet targets may trigger personal reprisals.

A highly centralized and bureaucratized system of funding and monitoring ensures accountability. In a continuum from the autonomy of the individual scientists in setting the research agenda to accountability to the government, science in totalitarian regimes is firmly at the accountability end (Nicholson 1977). The state controls the purse strings of both public and private funds. This control extends to foundations that provide grants to individuals or institutes, and even to the wealth of private individuals. In the United States, since the first decades of this century, such organizations as the Carnegie and Rockefeller foundations along with private universities have contributed significant sums to scientists and their institutes (Kohler 1991). In Soviet Russia, such organizations disappeared with the rise of Bolshevism and the nationalization of property. In Germany, the Notgemeinschaft der deutschen Wissenschaft, a federally funded foundation that saved a number of German scientists from financial ruin during the interwar Weimar Republic years, was subjugated to Nazi power as the Reich Research Council (Forman 1974), and all universities took their orders from the Reich Ministry of Education.

Since the state is the prime mover, its projects acquire significant momentum that carries beyond the completion of the initial goal. Bureaucracies everywhere seem to take on a life of their own, becoming institutions in search of a mission. But the centralization of science policy in totalitarian regimes enables one institute or a few to gain unassailable power to define scientific orthodoxy. Owing to this momentum, it is more difficult to derail economically unfeasible and environmentally dangerous projects than in pluralist regimes. In pluralist regimes, "dynamic orthodoxy"—that is, competition among researchers in university, industrial, and national laboratories for priority in discovery—ensures that scientific and public concerns are aired openly. Dynamic orthodoxy gives individual scientists, their institutions, and professional organizations greater autonomy in setting the research agenda, adjudicating the "facts," and considering public health, environmental, safety, and other normative concerns.

The second theme developed in chapter 4 is that large-scale technologies in totalitarian regimes acquire an aesthetic based on gigantomania. Public housing, subway systems, and government buildings have a depersonalizing scale. Their "ideological skins," for example, the neoclassical facing of Nazi government buildings, are thick, overpowering, and intimidating. The gigantic structures reflect the effort of officials and engineers alike publicly to demonstrate the strength, glory, and legitimacy of the regime, and as such they become symbols of the present and the future.

Proletarian Science and Aryan Science

The transformationist vision of the biological sciences, the ideologization of the physical sciences, and the primacy of the state in technology were all expressed in the notions of "proletarian science" in the Soviet Union and "Aryan science" in Nazi Germany. Proletarian science and Aryan science shared an essential belief that national science is the only true science. "True science" implied service to state-determined goals. Aryan science and proletarian science were superior to the science practiced by members of the international scientific community in terms of methodology, philosophical implications, and research emphasis. Both stressed applied research at the expense of basic science. Both justified autarky (self-sufficiency, but in this case international isolation) in science. Scientific contacts between scientists at home and those abroad were sharply restricted—for example, the exchange of scientific articles and participation in conferences abroad—because the Nazi and Soviet governments feared ideological contamination of national science.

Proletarian Science, Marx, and Stalin

To be sure, there were significant differences between proletarian and Aryan science. Proletarian science was a class-based doctrine. According to the Soviet version of Karl Marx's theory of history, historical materialism, society inevitably passes through a series of stages until its ultimate transformation into communism. These stages are slavery, feudalism, capitalism, and socialism. In a revolution or transformation from one stage of development to another, say from capitalism to socialism, the entire immense superstructure of philosophical, legal, and political ideas and institutions that arises upon a given economic basis also undergoes change. What happens to science? If science is part of the economic basis, then many of its salient features carry over to the next stage because of its implicitly cumulative nature. If science is a system of ideas and hence part of the superstructure, then it too changes radically, and socialist science differs greatly from capitalist science. Internationalist notions of science were pushed into the background (Josephson 1981). The superstructural conception gave rise to the notion of "proletarian science" as distinct from "bourgeois science," in which such phenomena as Lysenkoism prospered. Soviet proletarian science emphasized applications for broad social purposes. It ridiculed the theoretical orientation of bourgeois science as divorced from the needs of the masses and "ivory tower reasoning" of little purpose. The absence of the profit motive also distinguished proletarian science from bourgeois science. Soviet philosophers and ideologues debated whether science was superstructural or part of the economic basis, concluding only in the 1960s, when Leonid Brezhnev had become Soviet leader, that it was part of the basis. While under Brezhnev Soviet science became more international, strict controls over scientific contacts, attendance at conferences, and receipt of literature remained.

In addition to its own organizations and the KGB (secret police), the Communist Party organized a vast governmental bureaucracy to promote proletarian science within Soviet borders. These organizations monitored compliance with the ideological tenets of proletarian science. (The methodological uniqueness of proletarian science and its foundation in the Soviet philosophy of science, dialectical materialism, are discussed in chapter 3.) There were Party cells—groups of Party members meeting weekly and watching all activities—inside each research institute, industrial laboratory, and scientific association. The bureaucracy placed strict controls on access to Western scientific literature and contacts. Overt discrimination against such national minorities as Jews, Armenians, and Central Asians was widespread, especially concerning entry to universities and institutes or travel abroad. The treatment of such dissidents as the physicists Andrei Sakharov and Iuri Orlov indicates the extent to which the power of scientists was limited. (Orlov, long a believer in the human rights movement, became actively involved in opposition to Soviet policies precisely over the KGB's growing abuse of Sakharov in the early 1970s [Orlov 1991].)

Joseph Stalin gained power over the Communist Party and the Soviet Union in the 1930s, ruling until his death in 1953. He pushed breakneck industrialization and forced collectivization of agriculture on the Soviet Union in the 1930s. He set in motion the Great Terror in the late 1930s, as a result of which eight million citizens perished at the hands of the secret police. Stalin used all his sources of power to establish proletarian science. His government required planning of scientific research—what results could be expected and when?—to justify appropriations. Show trials were organized to condemn those who strayed from the newly established norms of scientific behavior. And by the end of the decade, the purges led to unimaginable damage to most scientific fields. The secret police were involved in the arrest of perhaps 10 percent of all physicists, 30 percent of all engineers, and most likely an equal number of biologists, many of whom perished in the Stalinist gulag (labor camps). But Stalin's proletarian science—applied, serving the masses, under central control, and ideologically pure—was firmly in place.

Aryan Science and the Führer (Leader) Principle

Hitler rose to power in 1933 during a parliamentary crisis. He had the assistance of conservative officials and right-wing military men who assumed they could control him to their ends. Quickly they learned that they were mistaken. The National Socialist Party used legal, illegal, and murderous tactics to destroy political opposition. Those who read Hitler's confessional autobiography and political diatribe, *Mein Kampf* (My struggle), knew quite well that he intended *völkisch* notions of racial purity and Aryan science to play a role in the Third Reich.

The natural unit of mankind in the German Reich was the *Volk*, a romanticized vision of the German peasant who, through organic ties to the soil, embodied the great German mission of constructing a worldwide civilization. The state existed to serve the *Volk*, a mission the Nazis believed that the Weimar Republic (1918–33), Germany's interwar experiment with parliamentary democracy, had betrayed (Gay 1968). All morality and truth was judged by its accordance with the interest and preservation of the *Volk*. The state reflected the "general will" of the people, so nothing in its laws could reject "*völkisch* tendencies."

Democratic government that relied on direct representation and universal suffrage could not succeed, since it assumed an equality within the *Volk* that did not exist. To a certain extent, scientists and engineers, like women, the working class, churches, and so on, gained access to power through the Führer. This was the "leader principle" that operated in all Nazi institutions and drew strength from the tradition of monarchic authoritarianism in Germany. In 1934, Hitler declared himself not only chancellor, but "leader." This meant he claimed not only constitutional powers but extragovernmental powers that required his followers to declare their allegiance to him. He expressed the true will of the *Volk*, so that any opposition or criticism was precluded. No interests or groups or ideas existed alongside him. "In place of conflicts and compromise, there was to be only the absolute enemy on whom the sights of the unified nation were fixed" (Bracher 1970, 340–44).

Since authority and power originated with Hitler, the fate of many projects depended upon him. Hitler supported Minister of Armaments Albert Speer's efforts to rebuild the center of Berlin as a monument to National Socialism and the expensive Nazi V-2 rocket program. But when scientists failed to get the führer's ear, their projects might languish, as the case of the Nazi atomic bomb project demonstrates. There was no question that Hitler intended the socially radical science of "racial hygiene" to achieve Aryan purity. He fully endorsed putting homosexuals, Gypsies, and Jews to death. He believed that the infirm (the sick, dying, and unfit) drained resources from the healthy and strong Aryan. German doctors and biologists willingly helped the führer achieve this end. Hitler believed that Germany must find *Lebensraum* (literally, living space) in the east, lands that would be freed from inferior Jewish, Gypsy, Russian, and Polish inhabitants by the *Wehrmacht* (war machine), with the survivors enslaved to create an agricultural work force to feed the German nation. German science assisted in meeting these goals through anthropological, geographical, and biological studies under the rubric of *Ostforschung* (research on the eastern lands).

What impact did the leader principle, *völkisch* ideas, and *Lebensraum* have on science? Like proletarian science, Aryan science promoted autarky. Aryan science served the nation, not profit or international Jewish capital. Aryan science was applied and technical, its supporters claimed, not overly mathematical, theoretical, and formalistic. Of course, true science originated among Aryan people. Philipp Lenard, Nobel Prize–winning experimentalist in physics and firm believer in Aryan science, wrote a history of "great men of science" to demonstrate that these men were related by blood, in Aryan kinship, as much as spiritually (Lenard 1933). Finally, the leader principle operated in science, so that Nazi Party functionaries often dominated the setting of the research agenda and the hiring and firing of personnel.

The Accommodation of Scientists to Proletarian and Aryan Science

If Soviet and National Socialist scientists were often on the cutting edge of international science, how did they respond to proletarian and Aryan science? In both Nazi Germany and the Soviet Union, scientists were forced to accommodate a new regime. The accommodation of scientific and technical specialists to totalitarian rule creates a tension when specialists promote economic, social, and political policies based on what they believe are the objective and rational methods of their engineering and science. Do experts derive political power naturally by virtue of their special knowledge? Technocracy means rule by technical specialists. Technocratic movements found fertile soil both in pluralist democracies and in totalitarian regimes. They were prominent in the former in the early 1930s as a response to the Depression; experts believed they could plan production and consumption more rationally than governments and could avoid the inefficiencies of market mechanisms. In Nazi Germany and the Soviet Union, party officials feared perceived technocratic trends even when there had been no attempt by technical specialists to pursue political power.

After the October 1917 revolution installed the Bolsheviks in power in Russia, most scientists avoided political involvement or playing an active role in serving the regime. A large number emigrated, several died from starvation and the difficult conditions brought about by war, revolution, and civil war, and others just wanted to be left alone to do their work, despising the Bolsheviks. But the academic community's precarious financial, physical, and psychological condition forced it into an uneasy alliance with the government.

Technocratic trends grew among members of the All-Russian Association of Engineers in the 1920s. Yet both from above (Stalin and the highest reaches of the Party) and from below (workers who resented the authority of specialists in a workers' state), opposition to the engineers grew pronounced. In response, Stalin's party apparatus promoted *praktiki*, who "mastered" technology in their day-to-day experience, to work alongside technically trained specialists, hoping that the two groups would converge in attitudes and responsibilities. The Soviet government was committed to the embrace of the most modern science and technology. It borrowed technology heavily from the West but strived to see that new social relationships developed around it. Conflict between specialists and workers was both class-based and grew from the tendency of the *praktiki* to damage equipment owing to their ignorance of modern equipment and methods (Bailes 1978). The Party also sought to replace so-called bourgeois specialists with scientists of proper working-class social origin and Marxian worldview. This effort was abandoned when it turned out that working-class "scientists" were poorly prepared to handle modern science; many had difficulty even with simple fractions.

The Communist Party subjugated the Soviet Academy of Sciences, whose prestigious institutes were the center of basic research, to its control in the late 1920s. Academy members long tried to defend their status and autonomy, in part by resisting Communist Party pressure to change the academy's charter or add new members. The sources of independence included a long history, tradition, and a secret ballot for membership that enabled individuals to vote from conscience, not because of political pressure. At this time, the Communist Party insisted upon adding new positions in the social and technological sciences to be filled with Marxist scholars and tip the balance of control of the academy to Communists, and from then on it sought to control the election process by approving slates of candidates beforehand. This tactic did not always succeed: the academy never removed Sakharov as a member even when he was persona non grata in government circles.

By the late 1920s, scientists were resigned to the fact that the Bolshevik government was there to stay. The scientists needed government funding at the same time as the government needed scientific expertise to achieve its many goals of industrial development, agricultural modernization, universal literacy and education, and public health. Even as the Communist Party increased its control over the scientific enterprise through ever more coercive means, most scientists had come to believe it was their patriotic duty to serve the nation, even if they disliked the Bolsheviks. In point of fact, their appropriations had increased, and they had yet to experience the dark days of Stalinism. Accommodation to the regime had occurred, even if the vast majority of scientists rejected the notion of proletarian science for a more internationalist view of science. Only around 8 percent had joined the Communist Party, and most of these individuals were social, not hard, scientists. Stalin's rise to power sealed the changed relationship between the scientist and the Party. He had put specialists on notice that technocratic aspirations would be crushed. They were fully accountable to the state. Stalinism was a further stage from accommodation to subservience to the state. In the long run, accommodation between the Soviet government and scientists was also achieved because of demographic factors: as the older scientists, the so-called tsarist remnants or bourgeois specialists, retired or died, young scientists trained entirely during the Soviet period replaced them.

In Germany, during the first years of Nazi rule, except for Jews and other minorities who were excluded from their professions by race laws, accommodation of science and engineering to National Socialism occurred as a natural part of the policy of coordination (*Gleichschaltung*) of all activities to Nazi principles of rule and discipline. Most scientists welcomed the strong, nationalist state. Aryan science and technology movements in chemistry, mathematics, physics, and psychology thrived spontaneously. It was not apparent immediately that these disciplines would be useful to the state; here scientists were initially vulnerable to political attacks on their professional standing that originated within their own ranks, from scientists or engineers who desired a more Aryan science. These attacks were not controlled by the Nazi hierarchy, and it eventually called an end to most of them, as we will see in the case of Johannes Stark and Aryan physics. The National Socialists did not deliberately set out to purge the scientific enterprise, only to remove Jews and other undesirables from their positions, an end accomplished automatically as part of the civil service rules of 1933. Such fields as eugenics and genetics neatly dovetailed with Aryan visions. Scientists in these fields quickly reached accommodation with the state (Renneberg and Walker 1994, 9–10).

While the Physics Society, the Kaiser Wilhelm Society, and similar institutions in Nazi Germany maintained a degree of autonomy, appointments were subject to racial laws and to the vagaries of National Socialist party politics and ministerial rivalries. The Nazi Teachers League and the Students League played a central role in seeing to it that Nazi dogma was considered in all facets of research. Yet scientists retained some autonomy in their academic organizations, academies, in matters of voting.

A complex relationship developed between technical experts and National Socialist power. The question was how to reconcile the international and rational elements of technocracy with the demands of the extremely nationalistic and often irrational Third Reich. German engineers had little political conscience and group identity, although they did have a national organization, the Union of German Engineers (Verein deutscher Ingenieure). The Nazis utilized their propaganda machine to manipulate engineers to support them by appealing to their sense of service to the nation and their desire to contribute to the glorious future. It did not matter that the Nazis used thuggish tactics. For engineers, "the opportunity to participate was honor enough." By the summer of 1933, six months after Hitler had become chancellor, all of the larger professional engineering associations had joined the Nazi-controlled Reich Society for Technological and Scientific Work.

There were many "Aryan" engineers who believed they could prove the racial superiority of Germans and solve the crises of bourgeois society as manifested in Weimar and the Depression. Aryan technology was anti-Semitic and hostile to the natural sciences, especially the new physics, for being too theoretical. Good Nazi engineers were expected to work for the *Volk*, and hence the führer as the historical embodiment of the *Volk*, but not to interfere in political and economic decision making. There was a short-lived technocratic movement in Nazi Germany. Its members published a journal, *Technokratie*. The movement paid lip service to Nazi ideology, but its members worried that Germany would revert to "primitive circumstances" if they followed the notions of Aryan science, and in fact followed the American and British movements closely. Then the Nazis closed the journal; there was room for technocrats, but not a technocratic movement (Heinemann-Grüder 1994; Renneberg and Walker 1994, 4–5).

In sum, engineers willingly worked for the Nazis on the development of armaments in exchange for such privileges as job security, good wages, adequate research budgets, exemption from military service, and the ability to serve the nation, if not the National Socialist German Labor Party. When the Nazi *Wehrmacht* and Hitler failed, the engineers believed it was not because of their mistakes but because of the failure of the Nazi system to utilize their achievements rationally and because of the haste, lack of planning, and political recklessness of the system. Still, German engineers had been willing to help the state achieve its irrational and thereby untechnocratic goals and policies. (Heinemann-Grüder 1994, 32–40).

Both Nazism and Stalinism led to a scientific diaspora. In the USSR, many biologists, physicists, chemists, and engineers, like ordinary citizens, simply disappeared into the gulag, never to return, and were granted only posthumous rehabilitation for their imagined crimes. Some served in slave labor camps put together specifically for scientists and engineers, the so-called Tupolev *sharashki* (after Andrei Tupolev, the great aviation engineer who fell into one of the camps captured so vividly in Aleksandr Solzhenitsyn's *The First Circle*). Other scientists were banished to the far ends of the Soviet empire to work anonymously in collective farms, car parks, or teaching positions that had little to do with their expertise. More biologists may have suffered from these fates than any other scientific specialists. In Nazi Germany, one in ten senior research biologists and professors lost their jobs and were forced to emigrate. Many lower-level researchers, Jews and leftists who were unable to get jobs in the West, would perish in the concentration camps. The Nazi diaspora and Soviet imprisonments were quite different, with one mainly racial and the other political in its roots. But in both systems, as in all countries where it has existed, totalitarian rule had profound and similar implications for scientific research and engineering.

Note

1. For the best treatment of the debates over what "totalitarian" means, see Gleason 1995. Gleason shows how the definition of totalitarianism has changed over the past fifty years in response to the rise of Stalinism, Italian facism, and Nazism, and the impact of Cold War thinking; it depends to a large extent upon whether the individuals who discuss it occupy the right, center, or left of the political spectrum. In this book, I do not wish to join these debates. But it is important to recognize that totalitarian regimes shape the face of science through their efforts to control, if not dominate, all competing power centers such as businesses, clubs, churches, or more generally interest groups.

Transformationist Visions: The Biological Sciences in Totalitarian Regimes

I N 1929 IN Tuskeegee, Macon County, Alabama, doctors working with the endorsement of the United States government Public Health Service commenced a study of syphilis in black males. Selecting more than four hundred individuals, the doctors observed the course of syphilis in these poor, rural residents over the next forty years. Alternately telling the men that they were not ill or giving them placebos, health service physicians allowed the syphilis to go untreated, even after an effective cure, penicillin, was discovered. Most of the men died of the effects of secondary and tertiary syphilis. The doctors justified this immoral medical experiment, and refused to acknowledge responsibility for it until 1972, in part because of a widely accepted view that syphilis was a "Negro disease" (Jones 1993).

The Western biomedical tradition has many other examples of the impact of cultural stereotypes and political belief systems on the conduct of science. Anthropologists, craniologists, and evolutionary biologists in the nineteenth century established a science of racial and sexual hierarchy in which white males stood at the pinnacle of intelligence. Their arguments were based on the size, shape, and various angular differences in human crania. Failing to find indisputable physical evidence of the "fact" of inferiority, they turned to intelligence tests as an allegedly objective measure of the inferiority of people of color. Scientists extended their studies to Slavs, Italians, and Jews (Gould 1981; Kevles 1985).

Around 1900, after the rediscovery of Gregor Mendel's paper showing that the occurrence of the visible alternative characteristics of plants is due to paired elementary units of heredity, now called genes, biologists set out to determine just what characteristics and behaviors are heritable. A mainstream offshoot of this program of research was eugenics.

15

Eugenists applied genetic knowledge for social engineering purposes, attempting to identify "feeble-minded," "sexually promiscuous," and other allegedly genetically inferior individuals by various scientific tests. Ultimately, their determinations were based on perfunctory observations of superficial physical characteristics or on performance in culturally bound IQ tests that allegedly measured "intelligence" by one mathematical standard.

The goals of eugenists had broad support, and many persons believed them to be desirable. Mainly, the eugenists sought to prevent the genetically inferior from reproducing, in order either to cut sharply the expense of caring for those considered unproductive if not worthless members of society or to "purify" the "race." In America, many state governments followed the eugenists' recommendations for policy. The measures included marriage laws and blood tests, segregation in camps for epileptics and the feeble-minded, and sterilization. Tens of thousands of "inferior" Americans were sterilized in the 1920s, 1930s, 1940s, and 1950s, lest they produce other "feeble-minded" offspring. Indiana first passed a sterilization law in 1907; twenty-nine states followed suit by 1930, with the blessings of a 1926 U.S. Supreme Court decision, in which Justice Holmes concluded, "The principle that sustains compulsory vaccination is broad enough to cover the cutting of fallopian tubes. Three generations of imbeciles are enough" (Kevles 1985; Vecoli 1960).

Support for using biology to transform society continued through the 1930s. In 1938, a U.S. doctor asked readers to consider the economic costs of supporting the "congenitally mindless" and incurably sick, and to euthanize them (Lennox 1938). Oblivious to Nazi horrors, in 1942 a New York medical doctor spoke for elimination of defectives: "So the place for euthanasia, I believe, is for the completely hopeless defective: nature's mistake; something we hustle out of sight, which should never have been seen in the first place" (Kennedy 1942). The publication of *The Bell Curve* (Murray and Hernnstein 1994), whose authors recommend overhauling various national welfare programs based on the lower performance of blacks than whites on IQ tests, indicates that the effort to base policies on a science of difference persists in the Western scientific tradition to this day.

Since the biological sciences in Nazi Germany and the Stalinist Soviet Union grew out of the Western tradition, they shared many of its assumptions. Yet their practice differed in both degree and kind from that in other systems. Eugenical practices limited to marriage, miscegenation, immigration, and sterilization laws in the United States and England, however immoral they were, were complemented in Nazi

Germany, first by official humiliation of Jews, Gypsies, and homosexuals, emblemizing their "inferiority" in the colored triangles they were forced to wear, and then by horrific sterilizations and maimings, euthanasia, and gassings of millions of people in concentration camps. This was done in the name of a superior, Aryan science with the complicity of German biologists and physicians who sought to end "lives not worth living."

In the Soviet Union, science of another sort developed owing to the confluence of Marxist ideology, Stalinist agricultural practices, and institutional centralization: Lysenkoism. According to Darwin's theory of natural and sexual selection, those organisms most fit to survive in a given environment, and to pass on their genetic material through reproduction, will thrive. Changing the environment will do nothing to change the genotype (genetically determined characteristics) of the species or of the individual; only random mutation creates individuals more fit for natural and sexual selection. As the biologist August Weissman demonstrated, cutting the tails off mice for any number of successive generations will not result in tail-less mice. This belied the older Larmarckian view of heredity, named after an eighteenth-century biologist who held that characteristics acquired during an individual's life could be passed on to descendants. By the 1930s, biologists everywhere had come to embrace the powerful combination of Mendelian genetics and Darwinian evolution. These concepts directed the research focus in embryology, plant hybridization, and animal husbandry and led to improvements in medical care, agricultural production, and so on.

Plant hybridization and animal husbandry were of central importance to Soviet planners and politicians who hoped to see a revolutionary new agriculture develop with the help of a revolutionary new Soviet biology. This biology, with proletarian methodology and focus, would permit agricultural production "to reach and surpass" that in the West. Unfortunately, mainstream geneticists could promise results only after careful study. At the same time, at Stalin's orders, Soviet agriculture was undergoing forced collectivization and grain requisitions. Peasants refused to enter the collective farms, slaughtering half the country's livestock rather than succumb to Soviet power. Famine ensued, particularly in Ukraine, where perhaps eight million persons perished.

In this environment, Trofim Denisovich Lysenko, a simple man of peasant background who rejected modern biological practices, secured complete authority. He promised the swift rejuvenation of collective farms with a proletarian science justified by hands-on practice that differed sharply from false genetic concepts. With the full weight of the Communist Party and Soviet government on his side, Lysenko required Soviet biologists to abandon Mendelian genetics and base their work on Lamarckian biology. Not until 1965 were biologists freed from this methodological straitjacket. The negative impact is felt in the former Soviet Union to this day.

There were great differences between Aryan racial hygiene and proletarian Lysenkoism. Racial hygiene embraced a strict biological determinism, the centerpiece of which was the belief that genes determined virtually all physical characteristics, intelligence, and behaviors of all living things. Lysenkoism was a diametrically opposed radical environmentalism, asserting that changes in environment would bring about changes in both the parent and its offspring. Yet Nazi and Soviet biology shared transformationist visions of changing society through the dedicated application of the science. In Germany, society would be transformed into a racially pure system. In the USSR, society and nature itself would be transformed into a socialist garden of material plenty run by workers allied to an all-powerful state. Direct and indirect political interference in the activities of scientists gave these transformationist visions their brutality and dominance.

The New Soviet Man and the New Soviet Biology

Before Lysenko rose to power, Soviet biologists developed a world-class research program that encompassed plant hybridization, animal husbandry, population studies, genetics, and eugenics. In Moscow, the tradition included N. K. Koltsov, founder of the Institute of Experimental Biology in Moscow, who played a major role in the development of genetics in the USSR. Like many Soviet scientific research institutes, Koltsov's had prerevolutionary roots in terms of organization and personnel. Like other scientists, Koltsov, a zoologist, saw the Bolshevik revolution as an opportunity to establish a multidisciplinary research institute. Koltsov secured modest government support to expand the experimental basis of his institute rapidly. His students and institute colleagues included A. S. Serebrovskii, S. S. Chetverikov, and N. W. Timofeeff-Ressovsky. Their studies of drosophila (fruit flies) contributed to a modern understanding of the evolutionary process, the notion of the gene pool, and the concept of "genetic drift" (Adams 1968, 1970, 1980).

In Leningrad, biological research thrived under Nikolai Vavilov, a plant geneticist, who was the major figure in the study of botanical populations. Vavilov studied under William Bateson, one of the founders of modern genetics, at Cambridge University and was a professor and president of the Lenin All-Union Academy of Agricultural Sciences (hereafter VASKhNIL), where Lysenko would later establish his orthodoxy. Vavilov traveled throughout the world collecting tens of thousands of specimens of wild plants and wheat for his studies on breeding and organized a series of agricultural experimental stations: near the cities of Leningrad, Murmansk, Vladivostok, Aktiubinsk, and in Uzbekistan, Turkmeniia, the Caucasus, Crimea, Ukraine, Belarus (Zhukovskii, Beliaev, and Alikhanian 1972). On the basis of his observations, he advanced the theory that the greatest genetic divergence in cultivated plant species could be found near the locale where each species originated. In Leningrad, Vavilov was joined by Iu. P. Filipchenko, the leading Soviet geneticist, who created several laboratories of genetics and eugenics that led to the creation of the Institute of Genetics of the Academy of Sciences under Vavilov in 1934 (Enkena 1962).

Koltsov founded the Russian Eugenics Society in 1921 and was editor of the *Russian Eugenics Review.* He considered eugenics an area of important research with potential medical applications, although he criticized the United States and Germany, where, he believed, sterilization and other laws were too quickly introduced on the basis of inadequate research. Other scientists shared Koltsov's concerns. Filipchenko tried to show that a Lamarckian view of genetics was no more progressive than the Mendelian was regressive. The Lamarckian view held out hope of social reform, since giving individuals a better environment would enable them to benefit during their lifetimes and to pass on the good effects of that better environment to subsequent generations. But this view was superficial and false, Filipchenko argued, because it assumed that only "good" environments had heritable effects. A consistent interpretation required "bad" environments to have an impact, too. And, if that were true, then all socially, physically, and intellectually deprived workers and peasants would have inherited the effects of having living for centuries under debilitating conditions. He rejected the conclusion that the Soviet working class was genetically weakened by the inheritance effects of its poverty and would not be able to rule for generations to come (Graham 1981, 231–45).

At the end of the 1920s, the golden days of Soviet biology suddenly ended. A period known as the Great Break signaled an end to the autonomy scientists had come to expect. The Great Break involved destruction of the better-off peasants as a class, rapid advancement of workers into positions of administrative responsibility, and class war against those considered to be remnants of the bourgeois past. Much of the impetus for the Great Break came from below, from militant young Communist workers who demanded that, at long last, they be given the opportunity to run this workers' state and its educational, cultural, and economic institutions.

Impetus for the Great Break also came from above. It coincided with Stalin's rise to power, his consolidation of control over the Party apparatus, and the introduction of economic policies that broke sharply with those of the past. One aspect of Stalinism was forced collectivization of the peasants into large-scale farms that were supposed to be more efficient and modern than anything the West had ever seen and would facilitate Communist Party control over the politically unreliable peasantry. Another feature of Stalinist economic policy was rapid industrialization, with an emphasis on heavy industry, not the consumer or housing sectors, let alone agriculture. The goal was to transform society into a giant socialist factory and promote economic self-sufficiency (autarky) in the face of what was called "hostile capitalist encirclement." Finally, the Great Break was accompanied by the introduction of Stalinist policies toward science, including the centralization of policy making in a few major bureaucratic organizations concerned more with the success of the industrialization effort than the health of science, the introduction of planning to control research from the level of the individual scientist to the institute, and the growing importance of Soviet ideological precepts in the conduct of research.

For biology, the implications of the Great Break and the introduction of Stalinist science policies were far-reaching. First of all, research, even that in the Academy of Sciences, became subordinate to the needs of collectivization through the Commissariat (later Ministry) of Agriculture. VASKhNIL ultimately gained approval from the commissariat to control the direction of agricultural research. Many observers argue that the strength of science in pluralist nations devolves in part from the decentralization of policy and competition among several research centers for scientific excellence. In the USSR, top-down centralization and the absence of competing scientific paradigms became the rule and permitted such individuals as Lysenko to dominate entire fields.

The growing autarky in the scientific establishment reinforced this negative control over researchers. World War I, the Russian Revolution (1917), and civil war (1918–21) had led to the physical and psychological isolation of Soviet scientists from Western developments; they had worked hard in the 1920s to reestablish international contacts. They relished the opportunity to travel abroad to conferences and to publish in German, French, British, and American journals. But Stalinist science policies required that they go it alone, to claim priority in scientific discovery for propaganda purposes and to avoid being tarred with the brush of "kowtowing" before Western science.

At the confluence of stunning advances by Soviet evolutionary biologists, rapid institutional growth, Stalinism, and the Great Break, Trofim

Denisovich Lysenko found fertile soil to achieve preeminence in the agricultural establishment. Lysenko, a mean-spirited, unaccomplished, pedestrian plant breeder, and his ideological mentor, I. I. Prezent, convinced Party leadership that they held the key to increased agricultural production. Lysenko believed in the inheritance of acquired characteristics and the primacy of the environment in bringing about changes in traits. Through such techniques as the "vernalization" of winter wheat or peas, that is, soaking seeds in water and keeping them cold or warm depending upon his whim, Lysenko believed he could manipulate the length of the vegetative period and turn winter wheat into spring wheat.

Vernalization was in fact a phenomenon that had been noticed in plants in the nineteenth century. It turns out that temperature change plays a role in the transition of a plant from one phase of growth to another. Winter grain crops that have germinated form spikes after exposure to cold. The technique works occasionally in shortening the vegetative and growing periods of winter wheat, but it does not involve an environmentally induced change of inherited characteristics. Still, Lysenko contended that it was possible to turn winter wheat into spring wheat given the right thermal conditions, and vice versa, and managed to establish vernalization as a revolutionary practice in Soviet agriculture.

Lysenko also applied various techniques to animals. He based his biology on Michurinist concepts, so called after I. V. Michurin, often referred to as the Russian Luther Burbank, a horticulturalist who selected plants and created hybrids, and who also accepted Lamarckian notions of the influence of the environment on heredity. In one case, Lysenko attempted to increase the milk butterfat content and raise milk production overall through cross-breeding purebred Jersey bulls, obtained at high cost from the West, with cows of other breeds. (Jersey cows have very high butterfat content in their milk, often 5 to 6 percent, but their yield is lower than that of many other breeds.) Knowledgeable animal husbandry specialists, armed with artificial insemination techniques, can breed such animals under controlled conditions. The problem is that while the offspring may have the desired productive qualities, they have lost much of their value for breeding. Their introduction into pedigreed herds will destroy traits produced over hundreds of years of breeding; high-butterfat milk lines would be weakened and eventually bred out. Lysenko hoped to counter the second-generation deleterious effects by selecting cows of large stature and "feeding them copiously during gestation which would force the embryo to maintain the desired high butterfat capabilities." But this was once again merely stressing environment over genes, and doomed to fail. Bulls in this line were used widely in state and collective farms, leading to a decline of herd quality throughout the USSR (Graham 1987, 45–46).

Since Lysenko believed in the inheritability of acquired characteristics, his science dovetailed nicely with the needs of the regime. The Bolsheviks intended to transform human nature radically in a short time with a new social structure and political system; Lysenko promised to create new plants and animals over a generation with changes in environment. Rather than waiting for genetics to bring about changes through lengthy study and evolution, he would revolutionize agriculture with peasants in the fields. Lysenko's promise of quick results and his devotion to the regime stood in stark contrast to the alleged "ivory tower theorizing" of drosophila counters and corn hybrid specialists. This was a case of "socialist" versus "bourgeois" science, lending credence to Lysenko's attack on scientists' allegiance to concepts of modern genetics and evolutionary biology (which he called Mendelism and Weissmanism depending on his whim).

Lysenko first gained public attention merely for being an example of the kind of success possible in revolutionary Soviet Russia for simple agronomists of peasant or proletarian background. By the mid-1930s, he had transformed attention from the media and upper echelons of the Communist Party into personal control over much of the biological research establishment. Suddenly, Koltsov's Institute of Experimental Biology was broken up into four different organizations, which lacked their former interdisciplinary strength; his colleague Chetverikov was arrested and sent into exile in the Arctic; and the research programs of the institute suffered large-scale interruptions. Koltsov resorted to attacks on fellow geneticists that sounded like Lysenko's in an effort to keep his institute operating.

Many scientists and institutes were not as fortunate as Koltsov. In 1936, Lysenko and his followers gained Party support to advance their program of Larmarckian biology. When some biologists, including Vavilov, attacked Lysenko both for his false science and for his failure to deliver a new, productive Soviet agriculture freed from the grasp of capricious nature and hostile capitalist encirclement, Lysenko relied upon Stalin's blessings to secure their removal. Lysenko was elected an academician, made director of VASKhNIL, and given Vavilov's post as director of the Institute of Genetics. Lysenko's teachings became dogma for the entire community of biologists. Modern scientific concepts fell from favor. In 1948, at a special VASKhNIL session, genetics was prohibited outright and reference to it removed from texts. Many geneticists were fired, exiled, and arrested. Some, like Vavilov, perished in the Stalinist gulag. In one of the more bizarre twists in the history of Soviet science, his brother, Sergei, became president of the Academy of Sciences in 1945.

What explains Lysenko's rise and staying power in the face of a tradition of pioneering genetics research and his own repeated agricultural failures? One explanation is the importance in the young Soviet republic of myths to demonstrate the superiority of the socialist system. Here was a peasant to be held up as an example of proletarian science, that is, of what was possible when an individual who was armed with Soviet materialist doctrines, and protected by the Communist Party, put his mind to the betterment of society. Lysenko relished the attention he gained and moved on to more bold pronouncements. These included the claim that Darwinian notions of the origin of species in Michunirist-oriented agribiology gave way to the discovery of the new path of the transition of one species into another species: for example, that warblers could give birth to cuckoos. He asserted that genotypical change was possible in a single leap, with retraining of plants through a series of plantings.

Some of his associates made equally bold discoveries. Olga Lepeshinskaia, an early associate of Lenin, said that cell membranes not cells themselves developed into diverse forms of life. This announcement rejected the discovery of German anatomist Rudolf Virchow that a cell can arise only by division from another cell. Here was a glorious, Soviet achievement: a law of the transition of nonliving material into living things. Lysenko also advanced the idea that evolution took place without competition among members of a species, a discovery based on experiments that demonstrated "mutual assistance" of complex plantings, on the premise that plants of the same species not only do not impede each other's development when sown closely together but actually promote the growth of their fellow plants. This finding of "socialism" among plants was clearly commensurate with Soviet socialism.

After Stalin's death in 1953, it appeared that Lysenko would fall rapidly from favor. V. N. Sukhachev, head of the forestry institute in Moscow, led the charge, but he was not alone. Biologists attacked Lysenko's theory of the transformation of one species into another and Olga Lepeshinskaia's claim to have created life out of Soviet ooze. They pointed to Lysenko's failure to use controls in experiments, the absence of rigorous description of experimental protocols, etc. At this time Khrushchev pushed the agricultural establishment to develop the "virgin lands" of western Siberia, Kazakhstan, and the Volga region into modern agriculture and to plant vast tracts of corn. Several new institutes opened, notably the Institute of Cytology and Genetics in the newly founded Siberian city of science, Akademgorodok, far from Lysenko's brown thumb. Physicists who needed data concerning the effects of radiation on living things, particularly in connection with their work on nuclear energy and bombs, established laboratories in their institutes to study radiation genetics. Once again, however, Lysenko was able to parlay the support of the Soviet leader into power over the genetics establishment until Khrushchev's ouster in 1964. Owing to his political fortunes, Lysenko always had greater resources in his hands— more equipment, more fertilizers, better strains or breeds to start with. This helped him to outrun, if not obscure, his essential failures.

In sum, the success of Lysenkoism can be explained by the fact that its biological notions conformed closely with the politically transformationist ideology of this activist state, and specifically with the concept of proletarian science. For Lysenko, the point was his promise of radical improvements in agricultural performance in the short term by embracing the practice of proletarian science, while the geneticists could only point to the potential of modern, scientifically based agriculture after a long period of experimentation. Further, Lysenko's science was interested not in large profits but in service to the masses. Indeed, in the United States, in the face of criticism that publicly supported research ought to benefit all members of society, a number of universities had meanwhile established "research corporations," which funded research from the profits generated by patents secured through the scientific efforts of their employees. Even if Soviet geneticists never embraced the profit aspect of bourgeois science, it was still possible to accuse them of being "bourgeois remnants" of the tsarist era. For example, Vavilov came from a well-to-do family many of whose members had strong contacts with European intellectuals. Finally, even the logic and methodology of proletarian and bourgeois science differed, as Lysenko's claims of anti-Darwinist and Larmarckian vernalization clearly demonstrated.

It mattered little that Lysenko's techniques failed to produce the desired results, let alone that they shared little of the replicability, controls, and rigor of bourgeois science. His biology enabled the state to introduce a new, purportedly scientific system for agriculture in the countryside, more noteworthy for extending Party control and organizing peasants into attentive laborers who increased yields ever so slightly than for its science.

Racial Hygiene in Nazi Germany

Nazi biology was also disaster, in this case for the millions of individuals who were sterilized, maimed, or gassed in the name of science. There are a number of common features in Soviet and Nazi biology that permitted their rapid growth: centralized administration of research, ideological and political interference, the presence of an activist state, and the promise of rapid results. Yet in one crucial way, the Nazi science of racial hygiene differed: it was a biologically deterministic doctrine in which heredity and genes mattered so much that the nurturist (environmental) concepts so prominent in the USSR played little role. Further, Nazi biology was not accepted by most non-Aryan biologists; it consisted of fantastic race theories and therefore was, like Lysenkoism, pseudo-science.

The German eugenics movement (and the Soviet one, for that matter) predated the rise of Nazism, commencing shortly after the ones in England and the United States. German eugenists debated whether "race hygiene" applied to one human race or different races (and hence connoted a hierarchy of racial worth), not whether it ought to have been rejected out of hand. While "eugenics" today is usually considered a pejorative term for a doctrine often equated with National Socialism, many of its early adherents came from the left of the political spectrum. These individuals noted that many eugenists opposed such traditionally conservative institutions of society as monarchy and aristocracy on the grounds that they were genetically regressive, allowing for no correlation between influence and natural ability. Many socialists believed that race hygiene concerned theoretical and practical measures for the improvement of race or avoiding its debasement. Some socialists argued that eugenics could be applied after class differences were eliminated from society, when the environmental and genetic causes of social ills could be separated. Leftist adherents, however, eventually objected to the increasingly racist sentiment among human heredity specialists.

The majority of eugenists tended to be members of the professional middle class, particularly medical men and academics from such fields as biology, zoology, and anthropology. They were practicing doctors, psychiatrists, lawyers, and others dealing with crime. These individuals felt threatened by a number of social developments, notably the creation of a mass urban society during the Weimar years. The historian Jeremy Noakes writes: "They both despised and feared the democratizing, levelling aspects of a mass society and what they saw as the crude materialism of the new elite." They were concerned with the decline in population growth in western Europe, which they attributed to a lower fertility of urban society. They worried about what they took as the degeneration of the race, which had resulted from declining birth rates particularly among the upper classes. They also believed that social welfare programs were "dysgenic" in that they interfered with nature's elimination of the unfit. The eugenists proposed keeping pedigrees of the gifted and the below average and passing measures to discourage dysgenic reproduction (marriage laws, confinement in asylums, sterilization) and to encourage the eugenically favored to have children (child benefits, tax reforms, propaganda) (Noakes 1984).

Legalization of sterilization for eugenic purposes was supported by the main association of eugenists, the German Society for Race Hygiene. The society grew rapidly: while it had only eight branches in 1929 by 1931 six new branches had formed and membership had approximately doubled. German racial hygienists had long advocated sterilization for race improvement. They had been fascinated by the efforts of the American eugenist Harry Laughlin to secure the sterilization of four hundred thousand individuals annually by 1980. (Such a large number of sterilizations, Laughlin contended, would be needed in that year unless more progress were made to control the genetically inferior by the 1930s.) The racial hygienists were also captivated by the studies of the Jukes family by Richard Dugdale and of the Kallikak family by H. H. Goddard, which were intended to prove that criminality, alcoholism, epilepsy, and prostitution were hereditary and which highlighted the cost to the state of prosecution, incarceration, and care. They were also intrigued with the U.S. Immigration Restriction Laws of 1924 that were directed against Jews, Slavs, and Italians, as well as antimiscegenation laws in the South.

Those on the right of the political spectrum often argued that science was value-free and that it was vital to follow the facts of human heredity to avoid genetic degeneration by giving advantages to help the strong and fit to propagate. They were particularly worried about the way World War I had led to the loss of large numbers of healthy young men and threatened to lower the birth rate further. The Nazi accession to power ensured that the racist aspects of eugenics retained prominence, while enabling the social consequences of a eugenical philosophy to be put into legal practice (Graham 1981, 223–31). The eugenists were given the opportunity to turn purportedly scientific study into political practice when Hitler came to power. In this way, the strong eugenics tradition in Weimar Germany was important for the development of Nazi doctrines of racial hygiene.

Given the intellectual commensurability between the racial ideology of Nazism and the biological determinism of the prevailing research paradigm, it is not surprising that quickly after the Nazi rise to power the medical professional was *gleichgeschaltet*, or synchronized, into a single, hierarchical structure with a vertical chain of command that was subordinated to party and state organizations. The National Socialists had the support of most scientists and engineers. The support was not active until the Nazis came to power, for scientists and engineers believed they must remain "apolitical" and above the fray. Yet they rarely opposed racial hygiene programs. Medical newspapers and journals were merged, closed, or subordinated to central command, while Jewish and other "racially inferior" physicians lost their jobs. Most scientists accepted these firings. New journals to popularize Nazi doctrines of racial hygiene were established and grew in readership and number rapidly (Proctor 1988, 71–77).

Medical men were among the earliest adherents of National Socialism. Many formed organizations intended to advance the doctrines of racial hygiene, science, and eugenics. The Nazi Physicians League, for example, grew to include 6 percent of the entire profession by 1933 (before Hitler's rise to power) and nearly a quarter by 1934. Many fewer engineers or judges joined the party, whereas perhaps 45 percent of doctors ultimately joined, with 26 percent of male doctors in the SA (*Sturmabteilung*, storm troopers or Brownshirts). Younger medical men were more likely to become party members, perhaps because they saw in National Socialism a doctrine of action (Proctor 1988, 65–67).

The fact that a biologically determinist doctrine fit well with National Socialist racial purity ideology contributed to great continuity in employment and funding of fundamental research in biology. The impact of forced emigration on biology was smaller than in other fields, with around 9 percent of individuals in the Kaiser Wilhelm institutes losing their jobs because they were non-Aryan. The Notgemeinschaft der Deutschen Wissenschaft (Emergency Association of German Scientists), a relatively autonomous body founded in 1920, was the most important German organization for funding fundamental biological research. In 1934, it was *gleichgeschaltet* under Johannes Stark, a Nobel Prize winner and "Aryan" physicist who imposed the führer principle on its administration to deny grants to Jews. In 1937, this organization became the Deutsche Forschungsgemeinschaft (German Research Association, DFG). The volume of government funding for the DFG grew many-fold in the 1930s and through the early war years. Party membership was not a prerequisite for DFG financial support for biologists, although it did not hurt (Deichmann and Muller-Hill 1994).

The impressive support for biological institutes after 1933 was due to politics, insofar as science was intended to contribute to the rise of the new Germany. The biologists willingly placed themselves at the disposal of the new regime. Even botanical research was well supported, as part of the political concept of "expansion of German living space" (*Erweiterung des deutschen Lebensraums*). Botanists emphasized the importance of their mutation research, particularly of research in polyploidy, for the fast breeding of new crop strains, most likely for use in conquered territories to the east. Researchers carried out their studies using slave workers in concentration camps. Most important for continuity in the funding of research was the fact that genetics was considered vital to Nazi race ideology. Even in schools, most biologists welcomed the mandatory introduction of "racial science" into biology curricula. Soon the schools were teaching how to safeguard the racial and genetic substrate of the *Volk* through genetics, race hygiene, population policy, genealogy, and ethnology (Deichmann and Muller-Hill 1994). The literature in primary and secondary schools was also intended to justify Nazi domination over alleged subhumans (Weiss 1994).

The Nazi state moved quickly to enact a series of laws directed at controlling the numbers of, and ultimately eliminating, individuals who were considered genetically inferior: as they came to be called, "lives not worth living." The first step was a sterilization law. Sterilization had been illegal in Germany until this time, although much of Scandinavia and the United States had eugenic sterilization laws and some German physicians had performed sterilizations without legal repercussions. A Nazi physician who had been active in the *völkisch* movement since the mid-1920s, Dr. Arthur Gütt, was put in charge of a committee that included other Nazi officials and sympathizers, which established guidelines by June 1933. Because of concern about the opposition of the Roman Catholic Church to this statute, which was passed in July, officials called it the Law for the Prevention of Genetically Diseased Offspring. Compulsory sterilization for a series of illnesses including epilepsy, congenital blindness or deafness, manic depressive insanity, schizophrenia, feeble-mindedness, Huntington's chorea, and alcoholism was now law (Noakes 1984).

By the end of 1934, courts of genetic doctors had already received over 84,000 applications for sterilization; they made over 64,000 decisions, with 56,000 in favor. Most individuals who appealed the decision were denied. By the middle of 1937, almost 200,000 persons had been sterilized, of whom 102,218 were men. Perhaps 400,000 individuals were sterilized during the Third Reich (versus 45,000 in the United States, which had a much larger population). This huge undertaking required enormous effort, extensive material resources, and new techniques, stimulating a "cottage industry" of doctoral theses on new ways to achieve sterilization cheaply, including tubal ligation and vasectomy techniques that were "refined" by the Nazis. One of the most horrific was the use of powerful X-rays focused on the genitalia of unsuspecting individuals who were filling out questionnaires at a special desk that hid the X-ray source (Proctor 1988, 95–117).

Nazi ideology, biological determinism, and totalitarian rule came together to allow eugenics in Germany to go much further. A German law permitting castration of "dangerous career criminals" and sex offenders was widely employed after the passage of a Law against Dangerous Habitual Criminals and Sex Offenders in November 1933. It was directed against homosexuals, even those suspected of being homosexuals, and against sex offenders on the flimsiest of evidence. Some twenty-two hundred were castrated between 1934 and 1941, with one prison doctor proudly boasting that he could complete the operation with only a local anaesthetic in eight minutes. The law and its enforcement reflected a view of sexual preference as wholly inborn, with heterosexuality essential to ensure the nativist call for a higher Aryan birthrate. During the war, the assault against homosexuals was extended to "failures, ne'er-do-wells, parasites, [and] good-for-nothings," with even "inclination" toward homosexual acts reason for castration (Giles 1992). The Nazi leader Joseph Goebbels was pleased that the führer decreed that those Nazi SS officers who violated article 175 against homosexuality would be punished by death. "This is a very wholesome decree which will render the elite organization of the party immune to this cancer" (Goebbels 1948, 63).

There was also a miscegenation law directed against Jews called the Blood Protection Law. There was even a special blood science, or serology, of the Aryan racial state. But in spite of the success of *völkisch* serology in mapping the distribution of supposed "races" as defined according to blood groups, those who targeted Gypsies, Jews, and Slavs for elimination did not have the time or interest to use blood measurements for the ends of racial purity (Mazumdar 1990). In reality, of course, neither blood type nor any other genetic characteristic can neatly separate, say, "Slavs" from "Aryans," owing to the huge genetic diversity within all human groups, so the program was a fantasy from the start. With advances in genetic typing it has become clear that there are no pure races, but only one species with many types of varieties.

Euthanasia, the next step in the Nazi racial hygiene program, had been widely discussed in the West since the late nineteenth century. There was a strong movement in support of it in Germany (Burleigh

1995). The goal of a euthanasia law would be to reduce the exorbitant costs to the state and society of maintaining those classified as defectives. In Nazi Germany, the euthanasia laws beginning in October 1939 were a dress rehearsal for the subsequent destruction of Jews, Gypsies, Slavs, homosexuals, Communists, and prisoners of war during the "final solution." Euthanasia began with children under three years old, then rapidly moved to include adults. How did this occur?

In August 1939, a Committee for the Scientific Treatment of Severe, Genetically Determined Illnesses produced a secret report, which was forwarded to all state governments, asking that midwives or doctors delivering any child born with congenital deformities such as "idiocy" or "Mongoloidism," missing limbs, spina bifida, and so on be registered with the authorities. Midwives were given an incentive to report these "lives not worth living": they were paid two marks for each registration. Soon, doctors were using morphine tablets, cyanide, and chemical warfare agents to kill perhaps five thousand of these children. The program expanded rapidly, with unofficial state endorsement, to children of six, then eight, twelve, and finally seventeen years of age.

Adult euthanasia followed. Doctors once again played an integral part in evaluating applications, as they did for sterilization. Many welcomed this program for furthering racial purity. They were paid five pfenning per evaluation if they reviewed more than thirty-five hundred per month, and ten if they evaluated fewer than five hundred per month. Of a total of 283,000 applications by October 1939, one-quarter (75,000) were marked to die. Carbon dioxide showers and crematoria were developed to kill dozens at a time. About one of every thousand German citizens was marked to be thus killed. By August 1941, 70,273 had been gassed. The eugenical elimination of those deemed inferior became a mundane occurrence; at a special ceremony held to mark the ten-thousandth cremation at one psychiatric institution, everyone received a bottle of beer. Nazi literature, art, films, even school textbooks glorified euthanasia (Proctor 1988, 177–92).

Beginning in 1942, the euthanasia effort shifted to injections, poisonings, and starvations to conserve resources for the fit engaged in the glorious war. Ultimately, historian Robert Proctor writes, this became less a single "reichwide operation and more the character of normal hospital routine. Doctors were never ordered to murder psychiatric patients and handicapped children, but were empowered to do so." Doctors had expressed some concern early on that their actions, strictly speaking, were not legal and in fact were never legalized. But the use of propaganda made it seem, to most of them, appropriate behavior to act on the public sentiment about a need to eliminate the unfit. This widespread sentiment ultimately created the right environment for the "final solution," the genocide of six million innocent people, to proceed. As Proctor concludes: "Each of these programs was seen as a step in a common program of racial purification. Medical journals used the expression 'life not worth living' to refer to those sterilized under the 1933 Sterilization Law, to those killed in psychiatric hospitals, and to those killed in concentration camps" (Proctor 1988, 221–22).

Nazi doctrines about the uselessness of these individuals permitted physicians to engage in horrific experiments in concentration camps. At Dachau, for example, Nazi doctors threw subjects into ice water to study hypothermia. They tested coagulants on prisoners' freshly cut amputations. They tortured other inmates with a decompression chamber. One physician gathered brains from individuals who had been gassed to test his theories on the connection between strictly physical characteristics of the brain and various illnesses, depressions, and diseases. Doctors injected people with gasoline, streptococci-laden pus, or tuberculosis bacilli. They carried out grotesque transplantations (Lifton 1986; Alexander 1949; Berger 1990; Advisory Committee 1994). Unfortunately, mainstream doctors rarely spoke up to criticize their peers who flourished in this pathological environment of Nazi ideology. To some degree, all German doctors embraced science to transform humans. Nature itself followed.

The Control of Nature in Nazi Germany and the Soviet Union

It was no coincidence that many Lysenkoist theories were advanced at the time when Stalin announced a plan for the transformation of capricious nature into a rational, orderly factory for the production of agricultural and forestry products. Initially, Stalinist planners had focused their attention on the urban landscape. But in search of economic autarky, desiring "natural" monuments to Stalin, and embracing the rationalist side of Soviet Marxism, the planners then turned their attention to nature itself. Under Stalin, transformationist "bulldozer" technology grew to epic proportions. All major scientific, engineering, and ministerial organizations were put to the task of transforming nature, on a scale never before imagined, into a smoothly operating machine to maximize its productive capabilities. This meant the development of natural resources, from timber to ore and fossil fuels, and harnessing waterways through a series of ever more grandiose canals and hydroelectric power stations (Josephson 1995).

Soviet engineers proposed transforming lakes, ponds, rivers, and forests into "natural" electrical energy and agricultural factories. Rivers would be "regulated" for irrigation and electricity, "narrowed and straightened, destroying the shoals"; ravines and marshes would be drained or filled, giving way to a system of orderly ponds and reservoirs. All aberrations from planned norms would be eliminated. A cascade of hydroelectric power stations, consisting of a massive unit touted as one of the largest in the world and several smaller ones downstream, would most efficiently take advantage of these rational, straightened rivers. Vast irrigation ditches would stretch to the horizon, with cheap electricity pumping the water to formerly arid, unproductive land. Then forest belts would be planted to protect the fields from the elements. One Soviet scholar asserted that complete mastery of nature was simply impossible under capitalism owing to anarchic distribution of property and to monopolies. A socialist order was required to ensure "complex rational utilization of resources." The problems of drought and "predatory economic relations" alike would be relegated to capitalism (Koshelev 1952, 2).

It was not always clear that this approach would predominate. In the 1920s, building upon a prerevolutionary tradition of conservation, ecologists and biologists moved to establish nature preserves, research stations, and university departments throughout the empire as a prerequisite for the rational use of the nation's great natural resources. But alternative approaches to industrial and natural resource development such as those offered by Vladimir Vernadskii and Peter Palchinskii, which urged greater circumspection and attention to the human and environmental costs, fell by the wayside. The eventual failure of the path of rational use is ironic in view of the alleged superior rationality of Soviet planning (Weiner 1988; Bailes 1990; Graham 1993).

The radical transformation of nature was the centerpiece of post–World War II reconstruction of ravaged Russia and a monument to Stalin. Dams were the big-ticket item. A series of government resolutions in 1950 called for the construction of the massive hydropower stations, with hundreds of miles of canals and hundreds of thousands of miles of irrigation channels. Planners intended each dam to lead to the introduction of hundreds, in some cases thousands, of square miles of irrigated land and reservoirs annually between 1952 and 1956, with a total area of 110,000 square miles (roughly the size of Nevada) put to new use in that brief period. [Koshelev 1952, 89, 92; Vinter 1951, 34–35, 57–65). According to the 1951 annual report of the major Soviet turbogenerator supplier, the hydropower stations, the canals, and the irrigation systems were "clear evidence of the unprecedented growth of our socialist economy, an expression of the outstanding success of the peaceful, constructive labor of the Soviet people." These huge projects were "the realization of the general line of the Communist Party and the Soviet government on the furthest industrialization of the country, of its electrification." The report concluded, "The fires that are being ignited on the great Russian rivers ever more clearly illuminate the path along which the great Stalin leads the Soviet people to Communism" (Velikie stroiki 1951).

Party officials and engineers seemed to ignore the great cost of these projects. On the human side, workers, many of them prisoners, were often subjected to brutal conditions. The infamous White Sea–Baltic (Belomor) Canal was built in the early 1930s by hundreds of thousands of slave laborers wielding picks, shovels, and wheelbarrows. Tens of thousands of deaths were justified by success in transforming enemies of the people into good socialists. The slogan of the laborers was "Prirodu nauchim—svobodu poluchim (We will instruct nature, and we will receive freedom)." In the book *Belomorsko-baltiiskii kanal*, the playright Maxim Gorky and his literary collective provided testimony of "the truth of socialism, of the brilliance of Stalin" and glorified the use of slave labor. Belomor symbolized "the brilliant victory of the collectively organized energy of the people over the elements of harsh nature of the north" and served as an example of "the mass transformation of the former enemies of the proletarian dictatorship and Soviet society into qualified collaborators of the working class and even enthusiasts of labor necessary to the government" (Gorky 1934).

These canal, dam, and irrigation projects also had significant environmental costs. They increased evaporation, salinity, and pollution of water resources and submerged productive farmland. In all, twenty-six hundred villages, 165 cities, and almost thirty thousand square miles (the area of Maryland, Delaware, Massachusetts, and New Jersey combined) were lost under the waters (Tushkin 1988, 8–13). Recent studies reveal that Soviet transformationist visions of nature will have a growing legacy of human and environmental destruction: high mortality rates resulted from industrialization without scientifically informed pollution controls on smelters and boilers; farmland has turned into desert; lakes have become saltbeds (Feshbach and Friendly 1992).

Two projects stand out for their transformationist hubris. The first involved the construction of paper mills on the shores of Lake Baikal; the second, the diversion of Siberian rivers. The Siberian river diversion project would have exceeded the pyramids as a monument of engineering prowess. All of the major Siberian rivers—the Ob, Irtysh, Enisei, and Lena—flow from the south to the north into the Arctic

Circle. Postwar engineers of nature planned to divert their flow through a series of canals, some reaching fifteen hundred miles in length, with associated hydropower stations and irrigation networks, to the arid regions of Soviet Central Asia (Gustafson 1981; Goldman 1972; Bressler 1992). Some far-sighted dreamers suggested using small thermonuclear explosions to melt snow, lift soil, move mountains, and dig trenches, all in the service of socialist geological engineering. Diversion plans gathered momentum in the late 1960s under Brezhnev. In all, nearly two hundred scientific research institutes, enterprises, administrations, and production organizations floated on the waters of this project (Voropaev and Ratkovich 1985).

The National Socialists hoped to tame nature, in distinction to the bulldozer-giantism of Stalin, from the opposite direction, by returning to nature. The Nazis did not succeed in putting science to the task of taming nature, but this was due to the fall of the Third Reich before engineers could undertake the endeavor. The historical roots of nature transformation were in place before the Nazi rise to power, in an irrational tradition that sought to link nature, antimodernism, and chauvinism. German environmental thinkers were repulsed by Americanization and the taming of the frontier. They linked love of home, homeland, and nature. They saw nature as cultivating important feelings: heroic purpose, community feeling, and devotion. One Nazi landscape planner observed, "Over and over again the love for plants and the landscape bursts forth from our blood, and the harder we try and the more seriously we search for reason [the more] we must come to realize the feeling we have towards a harmonic landscape and that the feeling of being related to the plants belongs to biological laws innate in our being." As historian Douglas Weiner asks: Why did Nazi conservationists not feel the same way about protecting their fellow humans? What explains the slide from nationalist nature protection to Nazi conservation? (Weiner 1992, 388–92).

The basis of Nazi thinking on nature, agriculture, and *Lebensraum* was the doctrine of "blood and soil," propounded by Richard Walter Darré, an academic who was appointed the Nazis' minister of agriculture by Hitler in June 1933. Darré, a specialist in agronomy and animal husbandry, gained Hitler's attention with his antiurban proposal to create new small and medium-sized peasant holdings to supplant the huge Junker estates in Prussia. Hitler found this plan reasonable. Heinrich Himmler's admiration of Darré led to the latter's appointment as chief of the Rasse- und Siedlungs-Hauptampt within the SS, the organization initially in charge of Himmler's *Lebensraum* resettlement plans. This was tied to efforts to carry out the "final solution"—

the genocide of the Jews and other unwanted peoples (Lane 1985, 153–55).

In 1928, Darré published an anthropological study of the relationship between *Volk*, soil cultivation, and the cultivator. He concluded there were two kinds of people, nomads and settlers. The settlers provided warriors and aristocrats and spawned the Nordic race. The nomads were the source of all other races, including Semitic, Oriental, and Indian. The nomads were materialistic, needing to concentrate on feeding and clothing themselves. They had no respect for property. They were communists. The settlers, on the other hand, recognized the essence of reality and the process of organic growth and changes, since the peasant, at least spiritually, remained on his property, tending his crops and animals. While the Nordic race could only be propagated on the soil, only nomads could successfully live and spread their cultural values in cities. These ideas were the foundation of the doctrine of blood and soil (Darré 1938; Lane 1985 153–157).

Darré set about late in 1933 to promote legislation in support of the Nazi program. The policies were intended to enable the *Volk* to become firmly rooted in the soil and bring about German racial regeneration. This required family farms capable of operating at higher than subsistence levels, and hence an end to the practice of dividing farms among several heirs. Clearly, the family and its lineage had to be protected. The true farmer had to be capable and racially pure. Since the *Volk* was the source of new, racially pure blood, the farmer had to abide by marriages only that guaranteed the health of the nation. The land had to be bequeathed undivided according to primogeniture and could not serve as loan collateral. The result was the Reich Inherited Freehold Act, which limited farm size to roughly three hundred acres. Yet fewer new farms came into being during the Third Reich than during the Weimar years, because land for military, industrial, and transportation needs had priority (Holt 1936, 178–208; Darré 1934).

Beyond this agricultural program, there were more sinister connections among views of nature, landscape management, and racist, exclusionist visions of utopia. Walter Schoenichen, director of the government Agency for the Preservation of the Monuments of Nature, believed that all Nazi youth should be taught by the end of the tenth grade the concepts of race, the measures necessary to preserve heredity, and natural history. The latter would be kept alive and developed by colonial adventures in Africa intended to yield feelings of desire for a revival of a German colonial Reich. Until that time, landscape planning had far-reaching importance for German settlement. As one

Nazi planner argued: "The interaction between men and the degree of cultivation of their land are of decisive importance for the development or destruction of mankind. Therefore [landscape planning] is the highest ethical commandment besides the Care of the Blood for us Germans. . . . For us National Socialists, planning results in the complete planning of space and economy, it aspires to the creation of a healthy structure of society and a permanent shaping of our living space as befitting to Teutonic German men." Preservation became the province of ultranationalists (Weiner 1992).

Agriculture, science, and nature were tied together in a paradoxical fashion through the notion of *Lebensraum*. National Socialism held fast to the romantic-racist peasant ideology, even though the idea of an agrarian state contradicted that of rapid industrialization, rearmament, and imperialism. The Reich Commissar for Consolidation of the German Nation was given power in 1939 by Hitler to deal with *Lebensraum* policies in the east. The goal was an empire stretching from Norway to the Alps, from the Atlantic Ocean to the Black Sea, linked by a gigantic network of highways lined with Germanic defense settlements to secure the territories. Expansion to the east via settlement thereby would provide strategic protection. Berlin would be a gigantic world capital like ancient Alexandria, Babylon, or Rome. Russia would be in part a vast army camp and in part an area in which Nordic "Reich peasants" racially selected according to a master plan were to be settled. The Slavic Russians would have to die out. They would be prohibited from procreating, and their schools would be closed to prevent development of an educated class. They were not "to learn more than at most the meaning of traffic signs." In case of difficulties, they could be rounded up into ghettos and liquidated by a handful of bombs (Bracher 1970, 407). This is precisely what happened to six million Jews in the "final solution." In this way, *Lebensraum* was partly an antimodernist battle for deurbanization through the establishment of Germanic farm settlements in the east. Yet the Germans did not reject the latest farm machinery and practices, so they were not antitechnology. They did oppose the large-scale approaches central to nature transformation under Stalin. National Socialism was not a rural movement, in spite of contributions of those like Darré to blood-and-soil ideology. Bracher writes, "The essence of the *Lebensraum* idea was the fusion of the national and imperial principles and the natural claim of the German race to territorial expansion and rule." (Bracher 1970, 288, 335–36).

Despite passage in 1935 of a Nature Protection Law, industrial development and militarization proceeded apace, since they were far more important to the Reich than conservation. Yet the prospect of leaving native German nature "wild" in the anticipation of gaining additional agricultural *Lebensraum* in the east soon loomed large. Under the Nazis, the fledgling policies of transformation found scientific verification in the field of *Ostforschung* (literally, research on the eastern lands). After the successful conclusion of *Blitzkrieg* (lightning war) on the eastern front, the evils that non-Aryan eastern peoples (Jews, Poles, Gypsies, and Russians) had inflicted on the land would be put to an end through their expulsion or elimination. Geographic *Ostforschung* combined racist population policy and scientific planning. A general plan for the legal, economic, and spatial bases for the development of the eastern territories was promulgated in 1940 and revised three times. The plan came out of the Planning and Soil Office (Stabshauptampt Planung und Boden) under Himmler, who was Reich Commissar for the Strengthening of Germandom from 1939. To ensure the organic growth of society based on rational principles of traffic, consumption, and administration, the plans outlined optimal distribution of people and goods, industry, and traffic lines. Each new *Hauptdorf,* or main village, would serve important administrative functions in the open spaces of the agrarian east and would unify administration in surrounding areas. There were thirty-six such settlements planned, with twenty thousand individuals each, organized in a linear fashion along railways and highways, but no cities.

Finally, the plan involved a program of selection, Germanization, and genocide. First the Poles, Jews, and Russian populations would be expelled and exterminated. Racial selection and control of the settlements of German peasants would follow. Then an "east wall" of military villages would be created. In the words of Mechtild Rössler, "Here the ideas of healthy, 'popular' design and construction, agrarian romanticism and anti-urbanism already developed in the Weimar Republic found their expression." There was racial planning and technocratic planning based on economic and technical categories. "There was Lebensraum in the East, and the purposeful, rational planning by the Reich Office" to enable a "silent transition" to post-war planning (Rössler 1994, 135).

Are Transformationist Visions Universal in the Biological Sciences?

A recent history of the U.S. Army Corps of Engineers indicates that this kind of extensive reworking of nature so prominent in Nazi Germany and the USSR was a major feature of the Enlightenment tradition, according to which nature could be studied, indeed ought to be studied, so that it would reveal its laws. Knowledge of these laws would enable men and women to apply science and technology to improve the quality of life of all members of society. American engineers as much as any adhered to a tradition, originally French, of centralized national planning, government financing, and military justification for projects (Shallat 1994).

What distinguished the USSR and Nazi Germany was the extent and centrality of the transformationist vision. Under Stalin, no obstacles to the transformation of citizens, cities, agriculture, and ultimately nature itself would be tolerated. This required a science that was radically environmentalist, a science that Lysenkoism wholly satisfied. In Nazi Germany, another biological pseudo-science served as the dominant research paradigm. In this case, it was a radically naturist science. This science, racial hygiene, meshed fully with state programs to promote the fantasy of a racially pure society. In both cases, the state underwrote the scientific research effort, in part through funding, in part through ideological support, in part by preventing other paradigms from being heard. The result in both cases was the destruction of life and nature as opponents to the regime, real or imagined, were identified and ruthlessly rooted out.

A major reason for the distortions that took place in the biological sciences in both Nazi Germany and the Soviet Union was the presence of an activist state at the top of which stood a charismatic dictator. Trofim Lysenko would never have risen to the top of the scientific establishment in the USSR without the help of Stalin, and later Khrushchev. At critical junctions in his career, when the tide of criticism seemed to be against him, the general secretary of the Communist Party intervened to protect him against so-called Weissmanist, bourgeois biological concepts. In Nazi Germany, similarly, state power was crucial in seeing the programs of racial hygiene come to fruition. As we shall see in the case of physics, too, ideology and centralized state control combined to create an enterprise that was subjected to inappropriate extrascientific controls and made researchers fully accountable to the state for their successes and failures.

The Physical Sciences under Totalitarian Regimes: The Ideologization of Science

T AKE AN APPLE, falling to the ground from a tree growing in Moscow, Berlin, London, or Cambridge, Massachusetts, in 1935. With certainty, you can predict the speed of the apple when it strikes the ground, if you know the height from which it falls. In fact, armed with Newton's laws—force equals mass times acceleration, for every action there is an equal and opposite reaction, inertia, and universal gravitation—you can determine earlier conditions or predict the behavior of objects larger than an atom and moving significantly slower than the speed of light. In comparison with the biological sciences, whose fate in totalitarian regimes we considered in chapter 2, it would seem that the objectivity of the laws of physics would render attempts by any government to regulate the practice of physicists fruitless.

Yet the practice of physics while that apple was falling was not that simple. Physics was becoming increasingly theoretical and, at the same time, more expensive to support because of its larger scale and complex apparatuses (Hiebert 1979). Physics was of strategic importance to every nation because of applications in communications, metallurgy, weaponry, and so on. Economic desiderata and political stuggles altered norms of professional behavior. All of these factors shaped the face of physics.

In the United States, the discipline had grown in the 1920s, including an increasing number of Jews who entered the field, among them I. I. Rabi, who would be America's fifth Nobel laureate in physics, and J. Robert Oppenheimer, future head of the atomic bomb ("Manhattan") project; some women also made a place for themselves. This was a time of tumultuous change in theoretical physics, but the United States lagged in this field, and, as they did everywhere else in the world, experimentalists predominated. Consequently many young theoreticians were forced to take postdoctoral fellowships abroad in Niels Bohr's institute in Copenhagen, Denmark, or with Max Born in Göttingen. For their part, the experimentalists encountered trouble gaining funding for increasingly expensive machines such as cyclotrons used to peer into the nucleus of the atom. Still, the discipline was healthy, fed in part by the immigration of such stellar scientists as Enrico Fermi, Hans Bethe, and Edward Teller, all of whom fled the rise of fascism in Europe (Kevles 1979). The United States became the center of world physics, as Germany had been before its expulsion of the Jews.

In Germany and the Soviet Union, too, physics underwent rapid change. Germany had a rich physics tradition; the USSR was a recent yet powerful entrant in the field. In both cases, political circumstances had ideologized the physics enterprise. This enabled some physicists and administrators to move outside such traditional channels of scientific communication as peer review and publication in refereed journals to employ bitter personal attacks. Instead of open and honest debates about ideas on the cutting edge of science, they used the power granted by state authority to push their ideas and frequently ruined the careers and even lives of physicists with rival theories.

In Germany, the basis of the ideological attacks was race. Germany was the birthplace of quantum mechanics. Berlin alone had seven Nobel Prize winners. But after the rise of Hitler, physics experienced significant losses. Most Jewish physicists were fired from their jobs; many emigrated for positions in the United States, England, or elsewhere. The modern physics of relativity theory came under attack for its counterintuitive mathematical formalism and seemingly Jewish roots. Even such a patriot as Werner Heisenberg could not avoid being tarred with the brush of being a "white Jew," and his career advancement was interrupted. Outside of the ideological sphere, the pressures of a war economy and intrigues between various National Socialist individuals and their bureaucracies had a negative impact on the conduct of research. The Nazi atomic bomb project would fail because of political, as much as scientific, obstacles.

In the USSR, the ideologization of physics had a class basis. To leading physicists such as Abram Ioffe, the founder of the preeminent Leningrad Physical Technical Institute in 1918, it seemed as if the laws of physics were under attack from malicious, if not ignorant, philosophers and physicists. Like their Nazi counterparts, they judged relativity theory and quantum mechanics to be mathematical formalisms, incapable of true description of nature. They criticized the failure of the new physics—quantum mechanics and relativity theory—to acknowledge somehow class conflict in their essence. The laws of physics in the twentieth century turned out to be universal within a narrow band of the physicist's activities.

Physics shared with biology striking symptoms of illness under totalitarian rule. In both Nazi Germany and the USSR, as a result of the ideologization of the norms of science, a kind of cultural revolution took place. Cultural revolution led to strict political controls over student admissions and faculty appointments. It fostered administrative interference in the professional life of the discipline—in peer review, publications, and financial awards. In both countries, the physicists' professional associations were subjugated to official party organizations. And there were grave personal losses. Hundreds of qualified scientists lost their jobs and emigrated; others were arrested, and some were executed. The extensive damage done to the physics enterprise in Soviet Russia and Germany was all the more surprising owing to the growing importance of physics in the twentieth century for governments at every point on the political spectrum. The most spectacular military achievements of midcentury—radar, jet engines, rockets and satellites, and nuclear weapons—are based on physical concepts.

In this chapter, by examining the experiences of Abram Ioffe in the Soviet Union and Werner Heisenberg in Germany, we will explore the major features of totalitarian physics. Ioffe was an experimentalist whose major direction of research was the physics of crystals; Heisenberg was a theoretician, one of the founders of quantum mechanics. Their experiences reveal the three major features that characterize the sciences in totalitarian regimes.

First, the economic interests of the regimes at once gave rise to pressures on the scientists to conduct research of an applied nature at the expense of basic science and, at the same time, created conditions that provided physicists with the intellectual ammunition—their research achievements—to defend their discipline from ideological intrusions.

Second, authority for science policy was centralized in bureaucracies that sought to harness research and development to state economic, political, and cultural programs. The Soviet Union and National Socialist Germany embraced central economic planning. Planners believed that their orders would compel research conducted in government institutes, universities, and factory laboratories to produce results in the short term with direct industrial and military effect. Their plans would secure economic self-sufficiency in the face of international political isolation. To a lesser degree in Germany than in the USSR, central planning led to the domination of entire fields by single individuals, research programs, and institutes. But the dynamism of competition between individuals and schools for scientific preeminence, a feature that German science had had in common with other countries like the United States, was dulled by skirmishes under the banner of Aryan physics.

Third, the essence of scientific autonomy came under outright attack. We take for granted that peer review ought to serve as the major avenue for determining whether research projects are funded or articles published. We assume that scientists know best the direction, content, scope, and pace of basic research. In totalitarian regimes, however, all aspects of research are carefully monitored, from the philosophical implications of scientists' discoveries to the activities of professional organizations. Science in totalitarian regimes requires the presence of both internal and external enemies as fodder for xenophobic ideological pronouncements, and it subjects perceived deviants from the norm to extraprofessional regulation.

Ioffe and Heisenberg wanted to be left alone to do their research. They assumed that physics was an apolitical endeavor. But they could not escape the political and social forces that swirled around them, nor the party functionaries who intended to control them. They had to make personal and professional sacrifices, some of which had significant moral and ethical consequences, to protect the physics discipline.

The New Physics

Twentieth-century physics presented a series of epistemological and scientific challenges to philosophers and scientists alike. In *Philosofiae naturalis principia mathematica* (usually referred to as the *Principia*) (1687) Newton set forth the notion of universal gravitation and three laws of motion (Dobbs and Jacob 1995). Over the next two centuries, physicists strived to incorporate all phenomena—heat and molecular motion, electricity, magnetism, light and other radiant energy—in one system. They built steadily and surely upon Newton's work, which clearly described how celestial and terrestrial phenomena such as planetary motions, the tides, and projectiles worked according to the same predictive laws. They made tremendous strides in this endeavor through the end of the nineteenth century.

Yet a series of phenomena fail to fit neatly into the Newtonian system. They include the physics of the very small (subatomic particles such as electrons and the alpha, beta, and gamma, particles emitted by radioactive atoms) and of the very fast (for example, visible light, ultraviolet light, and X-rays). Such scientists as Max Planck, who proposed that energy was distributed in a discrete, not continuous, form called "quanta," and Albert Einstein contributed to a revolution in physics that would give rise to relativity theory, quantum mechanics, nuclear physics, and astrophysics. In a paper published in 1905 on the electro-dynamics of moving bodies, Einstein overturned Newtonian concepts of absolute space and time serving as a universal frame of reference for all moving bodies. He promoted the "relativity principle" that re-moved the distinction between moving and stationary observers, based on the realization that all observers find that light travels at the same speed regardless of their own speed of motion or that of the source of light. His famous equation $E = mc^2$ revealed that energy had mass (more precisely, that electromagnetic radiation can carry inertia). In a paper published in 1916, Einstein incorporated gravity into his theory so that all cases of accelerated velocity, and also electromagnetic radiation, were included. This showed that space was curved as a function of the mass in it (Cassidy 1995). He predicted that light would be bent in passing the sun, just as the gravitational field around the sun bends the earth into its orbit around the sun.

Astounding developments in atomic theory meanwhile gave rise to quantum mechanics. Experiments confirmed the interrelation of con-tinuous and discrete phenomena such as light, which manifests wave and particle properties, and the existence of matter-energy. As physi-cist-historian Emilio Segrè points out: "According to all experiments on interference and diffraction, light consisted of electromagnetic waves. But according to Einstein's hypothesis, it was corpuscular in all energy exchanges with matter" (Segrè 1980, 150). Quantum mechanics required the synthesis of statistical and dynamic laws to describe the behavior of subatomic phenomena and pointed to the inherent difficulty in accounting for the interaction of the subject and the object in sub-atomic processes, including measurement itself.

This is the "uncertainty" principle. When we observe a macroscopic object, the perturbation in its behavior introduced by our observation is negligible. In the microworld, measurement, literally shedding light on the object, influences its behavior. We can know either its location or its momentum with complete precision, but not both at once. This uncertainty principle results from the fact that such an entity as an electron has a double nature, both wave and particle. Hence we use statistical or probabalistic descriptions such as the Austrian Erwin Schrödinger's wave mechanics or Werner Heisenberg's matrix mechanics to describe the particle's behavior with a great degree of confidence, but not with Newtonian certainty. Interestingly, the genesis of this change in physical thinking, which suggested that subatomic processes are acausal, was facilitated by an intellectual environment of despondency, pessimism about the future, and a tragic sense of loss of historic mission among German physicists (Forman 1971).

A second crucial concept of quantum mechanics was the notion of complementarity. Niels Bohr, whose 1913 model of the hydrogen atom was a foundation of quantum mechanics and whose Institute of Theoretical Physics in Copenhagen became a magnet for leading theoreticians during the 1920s, set forth this principle. Complementarity posits that the complete description of two apparently incompatible aspects of a situation must incorporate both without rejecting either: for example, the wave and particle aspects of light, or the Newtonian and quantum aspects of atomic systems. So significant were these changes that Bohr and Einstein debated the philosophical implications, with the latter struggling mightily but unsuccessfully against uncertainty and complementarity. Einstein's famous comment "God does not play dice" showed his distaste for the acausal quantum mechanics. His rejection of complementarity showed that Einstein was not a relativist in the philosophical sense; he believed as much as Newton in a single, absolute truth. Nor did he relish the increasingly mathematical nature of the development of the new physics.

Finally, the new physics involved new understanding of atomic structure, leading to nuclear physics. In 1932, James Chadwick reported discovery of a neutrally charged particle in the atomic nucleus, the neutron. Taken with the discovery of other particles, this enabled physicists to understand why the nucleus is stable in most circumstances but decays or undergoes fission in others. It explains the existence of isotopes, atoms that have the same atomic number but different weight: for example, deuterium, which has one more neutron than hydrogen. It remained to create artificial elements by bombarding atoms with neutron sources and to tame the power of nuclear fission in bombs and reactors. To achieve these ends physicists built particle accelerators, such as cyclotrons and Van de Graaff accelerators, and the world's first nuclear reactors, all by the mid-1940s. The physics of the very small and very fast also involved very expensive apparatus.

The new physics gave rise to epistemological paradoxes that confronted scientists everywhere. In Nazi Germany and the Soviet Union, those who rejected the new physics wondered out loud whether mathematical formalisms adequately described the real, physical world. Was theoretical physics an intellectual exercise that had little to do with reality? More critical were philosophical issues raised by relativity theory and quantum mechanics. Many physicists simply could not abandon Newtonian, mechanical explanations. They also confused physical relativity with philosophical relativism. But the physical theory of Einstein only overturned the privileged position of absolute space and time. Einstein said that space and time are conceivable only as relations among material objects and events and do not exist in a metaphysical sense in and of themselves. Einstein in no way asserted philosophical relativism; relativity theory does not deny the universal validity of physical principles or the existence of a single independent reality. Nevertheless it was disturbing that relativity required a rejection of so many classical notions, even the indestructible and unchanging atom and mass itself. These revolutionary changes in physics provoked lively debates in the revolutionary societies of the Soviet Union and Nazi Germany.

Soviet Physics: The Impact of Ideology and Administrative Control

Abram Ioffe (1880–1960), the dean of Soviet physics, lived through this tumultuous period. Ioffe was a representative of two worlds. He led efforts of physical scientists after the Russian Revolution to reach accommodation with the Bolshevik regime. Under his leadership, the Leningrad Physical Technical Institute (LFTI) was founded in 1918, and from it more than a dozen other institutes were created throughout the empire, often with personnel and equipment from LFTI. His own research focused on the physics of crystals, and he laid the ground for modern solid-state physics in the Soviet Union. His example provided other scientists with faith in their efforts to form government-funded research institutes in other fields. Ioffe worked with scientists in other fields to resurrect ties with Western science that had been broken by a decade of war, revolution, and civil war. He frequently traveled to Europe and the United States to purchase reagents, equipment, and journal subscriptions and to bring Western scientists up to date on academic affairs in the USSR. Ioffe led the Russian Association of Physicists to a decade of growth in membership, vitality, national influence, and international reputation, all by 1930.

Outside this professional and international world, Ioffe was expected to represent the Bolshevik regime and its pronouncements about workers' revolution and proletarian science. At home, he constantly faced skeptical Communist scientific administrators in his efforts to lobby for increased funding or physical plant expansion. Government encroachment on scientists' autonomy increased under Stalin. As for biology, so for physics Stalin's revolution from above, the so-called Great Break, marked a turning point in relations between scientists and the state. The changes involved centralization of administration, the introduction of long-range planning through five-year plans, and an emphasis on applied research for industrial production. Scientists' professional organizations lost their power, and international contacts were sharply curtailed. In spite of these changes, Ioffe and his colleagues insisted on believing that they could remain apolitical and ought to be left alone to do their research. They desired their science to be part of the international community and wanted their professional organizations, not the state, to determine the norms of their behavior. Hence there was no way they could avoid making political pronouncements concerning the relationship between science and Communist Party organs of power.

When the government introduced five-year plans for the economy as a whole, it also required rather detailed plans from scientific research institutes, ordering them to justify research and specify results. Administration was centralized under industrial and agricultural commissariats (ministries) and their financial and administrative organs. Symbolically, physics and chemistry institutes were transferred from the jurisdiction of the Commissariat of Enlightenment (education) to the Commissariat of Heavy Industry. Physicists initially provided quite detailed lists, itemizing between 150 and 200 individual projects for each year of the five-year plan. Such detailed plans constrained their activities, locking them into narrowly construed topics and making it difficult to embark on novel, unforeseen, and therefore unplanned topics of research. They complained about the onerous reporting requirements and thought it was dangerous to tie research too closely to narrow projects. At a major planning conference in Moscow in 1931, Ioffe urged Bolshevik administrators to recognize the danger of project-specific, long-range planning, which might constrain research and actually lead to a decline in productivity (*Nauchno-organizatsionnaia* 1980, 231–40). By the mid-1930s, scientists no longer had to provide comprehensive planning documents, having convinced officials that general plan outlines sufficed. But the pressure to do research commensurate with state goals of rapid industrialization did not abate, often at the expense of fundamental research (Josephson 1991, 141–62).

The Communist Party made an example of Ioffe's institute at a week-long session of the Academy of Sciences in March 1936, where physicists, Party dignitaries, and bureaucrats gathered to evaluate the performance of LFTI, its research program, and its director. The participants of the session were obligated to engage in the by now traditional Soviet "self-criticism" and to stress alleged failings of Soviet physics under the leadership of LFTI, such as the absence of a "science-production tie." They ignored the tremendous strides achieved under Ioffe's

direction: the formation of new institutes, the establishment of fundamental research programs, the training of cadres, and so on (*Izvestiia*
1936; Levshin 1936; Josephson 1991, 295–305).

Central planning did not have the desired effect, even with the added
pressure of such public criticism. But was it the physicists' fault? Planning was intended to prevent duplicate research efforts, but it often
prevented alternative research programs from developing and thereby
engendering healthy competition between scientific groups. Central
planning thus stalled funding for promising new areas such as nuclear
physics. Soviet leaders attempted to overcome this problem by intro-
ducing the concept of the "socialist competition" in all regions of the
economy, including the sciences. But it is artificial to plan competition, and the socialist competitions failed to produce the kind of drive
to be first in discovery or publication that often motivates scientists in
the West, where competition between various universities and national
laboratories exists more naturally.

Another way in which the Bolshevik regime infringed upon the autonomy of scientists was by subjugating virtually all professional organizations of physicists, chemists, and biologists (and architects and writers,
too). The Party feared a technocratic impulse. It intended to eliminate the last vestiges of independence that scientists, engineers, and
other specialists had achieved in national associations.

In the late summer of 1928, at the suggestion of Ioffe, the Russian
Association of Physicists gathered in Moscow, took a train to Nizhni
Novgorod, and then proceeded down the Volga River by steamship,
stopping at university towns along the way for popular lectures and
discussions. The major focus of the congress was the discussion of
Schrödinger's wave mechanics (Kravets 1928). But the boat ride down
the Volga, which had symbolized the independence of Soviet physicists, represented to an increasingly class-conscious Party these physicists' elitism and aloofness from the problem of building a workers'
state (Josephson 1991, 41–48, 130–39). At the next congress, in 1930,
physicists attempted to forestall encroachment on their autonomy. They
discussed how to accelerate training physicists of working-class origin
and Party membership. They spoke about the importance of planning
and of improving the tie between physics and production. They examined the question of the connection between physics and the Soviet
philosophy of science, dialectical materialism (covered in the next section). But soon after, the Russian Association of Physicists disappeared
without a trace. In its place arose several advisory boards and planning organizations within the Commissariat of Heavy Industry and the
Academy of Sciences. While Ioffe and his colleagues served on these

boards, which represented all scientific fields, the new organizational relationship symbolized their accountability to state industrialization programs (Josephson 1991, 141–62).

Cultural revolution, which accompanied the Stalinist revolution from above, was intended to lead to the replacement of bourgeois specialists with scientists of the proper social origin and worldview. The Party mobilized society for the industrialization effort through highly publicized show trials in the late 1920s in which some scientists and engineers were convicted for "wrecking" state industrialization plans. The Party used "wrecking" as a pretext to advance Communist cadres into the ranks of scientists through cooptation, coercion, and *vydvizhenie*.

Vydvizhenie was the state-sponsored advancement of workers into positions of responsibility in economic enterprises, higher educational institutions, and scientific research institutes on the basis of their class origin and Communist Party affiliation rather than on merit, qualifications, or other traditional reasons for advancement. Cultural revolution therefore was a summons for a Marxist cleansing of the R and D apparatus, from ideology to personnel. *Vydvizhenie* required universities and institutes to pay strict attention to the social and religious backgrounds of their students and employees and to ask if they were good Communists who embraced dialectical materialism. *Vydvizhenie* reflected a massive effort to train thousands of scientists and engineers to build up Soviet industry and agriculture.

Yet the Party had limited success in its effort to transform the exact sciences. (It was much more successful in the social sciences, which were already a Marxist stronghold.) *Vydvizhenie* failed in the physical sciences for several reasons. It was challenging to train even the most gifted student in quantum mechanics within the time constraints set by the cultural revolution. Why would it be different with proletarians, who often lacked such rudimentary skills as the ability to do fractions? Leading scientists such as Ioffe urged caution in carrying out *vydvizhenie*. Ioffe acknowledged that the country needed a huge army of sociopolitically conscious and steadfast workers, but he worried that the most important qualification was a natural gift for scientific creativity (Ioffe 1930, 26; Josephson 1991, 184–203). The Party scaled back the radical personnel goals of cultural revolution by 1934 because it had been a dismal failure.

While in terms of personnel advancement its record was mixed, the Communist Party was far more successful in another aspect of the ideologization of science, that of drawing a sharp distinction between Soviet, proletarian science and capitalist science. It achieved this distinction in several ways. One was the promotion of autarky in science.

Soviet physicists had been world leaders in the development of quantum mechanics and relativity theory. At Ioffe's institute Ia. I. Frenkel', V. F. Frederiks, and A. A. Friedmann not only contributed theoretical advances but were the leading early popularizers of relativity theory in Soviet Russia. (Frederiks and Friedmann 1924; Frenkel 1923). Similarly, Soviet physicists rapidly embraced quantum mechanics and debated its epistemological implications and physical content (Ioffe 1927). Now, in the 1930s, they were required to go it alone, divorced from developments in international science. The Party tried to limit contact with Western scholars and ideas. This was the logical outcome of proletarian science's denigration of the achievements of Western science and of Stalin's contention that he would create "socialism in one country." The pressure for autarky in science caught scientists between countervailing forces. Scientists like Ioffe had expended great effort to reestablish ties with the West and recognized the importance of those ties for the health of the scientific enterprise. The Party sharply curtailed contacts in both directions after a number of leading scholars went abroad and failed to return: for example, George Gamow, author of the "big bang" theory of the origin of the universe in 1948, who never came back from a conference in 1933 (Gamow 1970, 110–23).

In 1934, Stalin refused to permit Peter Kapitsa, a future Nobel Prize winner who had worked in Cambridge, England, since 1921 but spent the summers in the USSR, to return to England. Through the intervention of his mentor at Cambridge, Ernest Rutherford, and of Niels Bohr and others, the Soviet government agreed to purchase Kapitsa's Cambridge equipment and build a new institute for it in Moscow. At first Kapitsa refused to work, stunned by his detention and shoddy conditions for research. His letters and phone calls of protest to government officials went unanswered. He was followed by the secret police. He was isolated and alone. He began to criticize the "narrow and utilitarian" approach of the government to science and likened himself to a violin being used as a hammer. Kapitsa remained courageous throughout, writing letters to Stalin himself on behalf of imprisoned colleagues and criticizing censorship and ideological interference. But autarky in science had been established. Kapitsa wrote, "It is all very sad indeed that the political conditions are such that the country which claims to be the most international in the world actually places her citizens in such a position that it is very difficult for them to visit other countries" (Boag, Rubinin, and Shoenberg 1990, 317–40; Badash 1985; Kapitsa 1989, 27–61).

The number of Soviet scientists who traveled to the West now dropped precipitously, and although exact numbers are not available, it seems they could be counted on a few hands until after the death of Stalin in 1953. The centralized receipt and censorship of scientific publications also grew into an art at this time. Physics institutes continued to subscribe to and receive a large number of Western scientific journals, but scientific information was allowed to diffuse only very slowly while journals were cleansed of any "anti-Soviet" information. Even requesting or receiving a reprint was a serious matter (Josephson 1991, 303–8).

Another way in which the Party turned up the pressure on scientists to conform to Soviet norms was the ideologization of science, which had both organizational and epistemological components. Party officials and Stalinist ideologues established study circles on Marxist theory in research institutes. In the circles scientists and philosophers debated the fundamental tenets of dialectical materialism, historical materialism, and Party history. Assisted by Marxist scientists, they decided that certain fields of modern science—genetics, relativity, and quantum mechanics—were "idealist" or pseudo-science and hence harmful to the proletariat. They insisted that scientists recognize their research activities as inherently political. "Ivory tower reasoning" would have to give way to science for rapid industrialization.

Ultimately there were several hundred of these circles in which leading scientists participated, including national organizations like the Circle of Physicist-Mathematicians Materialists with chapters in places as far-flung as Baku (Azerbaijan), Tbilisi (Georgia), and Kharkov (Ukraine). Others were methodological seminars that met regularly at such institutes as LFTI (A AN, f. 351). For example, in December 1930 in the study circle at Moscow University's physics department, Ioffe spoke about socialist competitions, "shock work" (all-out surges of effort), the dangers of planning and *vydvizhenie* for science, and the philosophy of science (A AN. f. 351, op. 2. ed. khr. 26). Boris Gessen chaired a number of the Moscow sessions and presented a series of papers at other sessions. What were the major issues that ideologues and physicists contested?

The Ideologization of Physics

Dialectical materialism is based on three principles: "All that exists is real; this real world consists of matter-energy; and this matter-energy develops in accordance with universal regularities or laws." It has nineteenth-century Russian roots, although it grew out of the writings of Marx and Engels. Since Marx and Engels provided only general outlines of their view of the relationship between the materialist philosophy of nature and modern science, frequently in notebooks and un-published essays, it remained for Soviet writers to clarify the details (Graham 1987, 24–67).

One of the details was how Engels's three dialectical laws of nature (the interpenetration of opposites; the negation of the negation; and the transformation of qualitative changes into quantitative ones, and vice versa) would be applied to understanding the new physics. An example of the first law might be a magnet with a north and south pole, or the wave-particle duality of light. Not all of the philosophical disputes of the 1930s so neatly reduced to a consideration of the ap-plicability of these laws to modern science. But many of the partici-pants in the debates believed that they did.

The representatives of the two major trends of Marxist philosophy, the "Deborinites" and "Mechanists," organized a number of research institutes in the mid-1920s with a triple purpose: they wished to be-come more conversant with recent advances in the sciences; they in-tended to train *vydvizhentsy*; and they desired to attract natural scientists to their fold. The Deborinites believed that the epistemological ques-tions that had arisen in response to the major developments in physics in the first third of the century—the rise of statistical physics, the wave-particle nature of matter-energy, and indeterminacy at the micro level—demonstrated the compatibility of modern physics with dialectical materialism (Joravsky 1961, 279–97).[1]

The Mechanists took exception to many of the Deborinites' posi-tions. They believed that all processes in the external world could be explained in terms of the laws of classical mechanics. They referred to the works of leading Marxist scholars, Engels and Lenin in particular, on mathematics, physics, chemistry, and biology to demonstrate that mechanical processes in both the organic and inorganic worlds reduce to matter in motion, subject to the concept that all qualitative differ-ences are differences of quantity (Ot redaktsii).

Many Marxist philosophers contested the "idealism" inherent in quan-tum mechanics and the "philosophical relativism" they believed per-vaded Einstein's work. They criticized the physicists on several grounds: for accepting the indeterminacy principle of Heisenberg; employing mathematical formalisms, which seemed to serve as a substitute for, if not deny the existence of, physical reality; giving in to "idealist" un-derstandings of probability in the wave mechanics of Schrödinger; con-tending that electromagnetic forces acted at a distance; and allowing the disappearance of mechanical causality from modern physics. Hence Soviet researchers were frequently required to adapt scientific concepts to Marxian ones.

This ideologization of science permitted individuals to advance their careers outside of the normal scientific channels. A. K. Timiriazev (1880–1955), son of the famous agronomist and biologist K. A. Timiriazev, a professor of physics at Moscow State University and a Party member from 1921, was prominent in this regard. Not an outstanding scientist, Timiriazev was distinguished by his unceasing devotion to classical physics, a strict allegiance expressed in everything he published. Timiriazev succeeded in turning the Moscow University physics department into a kind of armed ideological camp. He tried to force graduate students and faculty alike into his circle of militant mechanist physicists, reserving his most hostile commentary for theoreticians, in particular such Jewish ones as L. I. Mandelshtam, Ia. I. Frenkel' and L. D. Landau. Timiriazev's hostility to Einstein and relativity physics took on such proportions that he sarcastically denied in a public meeting having suggested that Einstein be shot (A MGU; *Mekhanisticheskoe* 1925).

Timiriazev opposed the rejection of Newtonian mechanics required by relativity theory to explain the behavior of elementary particles and radiant energy. Like Philipp Lenard in Germany, Timiriazev posited an ether that filled the universe to explain the transmission of electromagnetic energy mechanically through space. Like other so-called Mechanists, Timiriazev was troubled by the increasing role of statistical laws, the "mathematicization" of matter, the apparent rejection of causality, and so on.

Boris Gessen, a professor of physics at Moscow State University and a Marxist, stood at the forefront of the Deborinite defense of the new physics. Gessen's defense of quantum mechanics addressed many of the philosophical and physical issues raised by the development of statistical laws, the wave-particle duality of matter-energy, the uncertainty principle, and the inherent difficulty of accounting for the interaction of the subject and the object in subatomic processes, including measurement. In each case, Gessen concluded that quantum mechanics and dialectical materialism were commensurate. The dialectical law of the interpenetration of opposites was reflected in the wave-particle nature of light; the existence of matter-energy, and the vital new understanding of the relationship between the subject and the object. Gessen pointed out that the new physics permitted abandonment of the concepts of absolute space and time, which had taken on metaphysical significance in classical physics (Gessen 1930).

In 1931, at the first international congress in the history of science in London, Gessen delivered a paper entitled "The Social and Economic Roots of Newton's *Principia*." In it he described how the economic forces of nascent capitalism, the political liberalism of the rising middle class, and the religious doctrines of the English church had made likely the appearance of someone like Isaac Newton to codify laws of physics. These laws were important for mining and metallurgy, ballistics and trajectories for the military, and other aspects of British capitalism. It was unimportant that Newton was a genius. Under the action of social and political forces, Gessen argued, some individual soon would have come along to discover those laws (Gessen 1971).

In this paper, unlike any other he ever wrote, Boris Gessen had used the language of basis-superstructure theory (in which ideas were a superstructure resting on an economic basis) and proletarian science. Simultaneously, he employed Aesopian language to defend relativity theory and quantum mechanics—for these had come under attack at home in the Soviet Union as worthless "idealism" (Graham 1985). Behind his contention that the value of Newton's physics was not diminished by the fact that he believed in God and was a product of the bourgeoisie was Gessen's unspoken argument that although Einstein was a Jew, allegedly a follower of an empiricist, subjective, idealist philosophy akin to that of Ernst Mach, and a product of bourgeois society, all this did not require rejection of relativity theory, a valid scientific concept. In spite of the best efforts of Gessen, and of dozens of more competent physicists than he, the entire physics enterprise fell under attack from Marxist philosophers of science of various stripes and would remain a philosophical battleground until after Stalin's death in 1953.

In a series of increasingly bitter public disputes in the mid-1930s, a number of Timiriazev's allies engaged Gassen, Ioffe, and other physicists over the problem of the weightless, imponderable, and unknowable ether to explain how electromagnetic forces act at distance. Mainstream physicists concluded that the Mechanists' hypotheses were anachronistic and based on shoddy work. Timiriazev appealed to a Party functionary, V. M. Molotov, chairman of the council of ministers, for redress. As in the case of Lysenko, it appeared that there would now be a special session to air this dispute, with unpredictable, but certainly damaging, results to physics. Fortunately, Timiriazev could not rely on the support of Marxist philosophers, few of whom could handle the epistemological complexities raised by the new physics. In addition, he faced nearly universal condemnation from the physicists, who, in addition to well-reasoned philosophical rebuttals, insisted that the proof of the new physics was in the many significant Soviet achievements in electrification, communications, and metallurgy (Josephson 1991, 228–32, 252–61).

Yet vindictive disputes spread from public forums to political and scientific journals. Physicists were called out, shouted down, accused of treason, and run out of town. Ideologues and administrators had gained the upper hand in scientific disputes over the philosophical content of physical theories. Ioffe himself now courageously stepped forward. In a major article published in the leading theoretical journal of the Party in 1937, he took issue with both individuals and tactics. He questioned the traditional Bolshevik rule of disputation that "he who is not with us is against us." He demonstrated the fallacy of continued adherence to a Newtonian view of electromagnetic phenomena. He concluded by lumping together the antirelativists in the USSR with their Nazi "allies," Stark and Lenard, calling them reactionary and anti-Semitic (Ioffe 1937).

These seemingly abstract debates had a deadly context. In the mid-1930s, the Great Terror spread throughout Soviet society, and it hit scientists hard. Stalin set the Terror in motion, personally signing lists that condemned hundreds of thousands of innocents to death. The Terror gained momentum in 1937–38, moving well beyond such visible enemies as the Old Bolsheviks, who were potential rivals, to envelop all of society. Most likely ten million died; ten to fifteen million were interned at one point or another in Stalin's labor camp system.

In 1937, the assault on physics left the editorial boardroom and entered the physics community at large. Dozens of leading scholars were arrested; hundreds of midlevel physicists, too, lost their careers or lives. The Terror struck the entire discipline, in Moscow, Kharkov, Dnepropetrovsk, and above all in Leningrad with its leading theoreticians. Theoreticians Vladimir Fock and future Nobel laureate Lev Landau spent time in prison. Gessen was arrested and shot, along with such promising young physicists as Matvei Bronshtein (Gorelik and Frenkel 1994). Perhaps 10 percent of all physicists perished (Kosarev 1993; Josephson 1991, 308–17; Conquest 1968). The repression in astrophysics and astronomy was even more devastating than in physics, with the discipline nearly destroyed. The rational science of central planning had given way to the xenophobic, irrational proletarian science. Even Ioffe could be heard at a conference on low-temperature physics in Kharkov in January 1937, expressing "the anger and indignation of Soviet scholars at the ignoble work of Trotskyite bandits which demands from the proletarian court its destruction" (*Zhurnal tekhnicheskoi fiziki* 1937).

In June 1941, although he had signed a nonaggression pact with Stalin, Hitler unleashed a surprise attack on the USSR. His armies quickly conquered much of the European USSR. More than twenty million Soviet citizens perished in freeing their country from the invasion. World War II and the postwar effort to rebuild the devastated country momentarily diverted the attention of Party officials from the ideological sphere.

Moscow University physicists were able again to raise the specter of "idealism" in physics in connection with Cold War fears. In 1947, a renewed effort to criticize "cosmopolitanism" (Western sympathies) commenced under the name of the Zhdanovshchina, involving unrelenting pressure on scientists, artists, writers, and musicians to conform to the Stalinist standard and to avoid "kowtowing" before Western ideas. Between November 1948 and May 1949, physicists centered at Moscow University called a series of national meetings to renew the attack on quantum mechanics and relativity theory and their Soviet representatives. With the blessings of the Communist Party, they set out to organize a major conference in physics like the one in biology that gave Lysenko unquestioned authority. Fortunately, physicists working on the atomic bomb project got wind of the idea, called Lavrenti Beria, head of the secret police, and informed him that a bomb could not be constructed without taking note of relativity theory and the equivalence of matter and energy. The conference was not convened (A AN f. 596).

But grim ideological pressures remained. They were sufficient to force Ioffe, the father and protector of Soviet physics, into actions that appear to the outsider to be uncharacteristic for him. At a March 1949 meeting in his institute, Ioffe read a speech noteworthy for its Cold War mentality. He called for open struggle in the military and ideological spheres against enemies at home and abroad and even attacked a close, lifelong colleague, Iakov Frenkel', for alleged ideological failings (A LFTI) Ioffe, too, then fell to such criticism in the early 1950s and was removed from the directorship of the institute he had founded. Physicists would not reassert control over philosophical issues until several years after Stalin's death. The Soviet system of centralized control, autarky, and planning would handicap the performance of Soviet science until the regime collapsed, and indeed to this day.

In spite of all this, physicists were able to maintain a modicum of authority. They participated in the genesis of quantum mechanics and remained current, even if their government now prevented them from traveling abroad to conferences or even exchanging reprints without difficulty. The industrialization effort ensured that budgetary allocations grew significantly. The importance of physics for communications, electrification, metallurgy, and other industrial programs provided a shield to physicists. The financial support given to theoretical departments was tacit acknowledgment by the authorities of the validity of theoretical endeavors, so long as they were accompanied by applied pursuits. Through participation in Marxist study circles, physicists were prepared, if reluctant, to discuss epistemological issues from a position of strength. The complexity of philosophical issues surrounding relativity, quantum mechanics, and nuclear and solid-state physics also made it difficult for ideologues to challenge them. In the end, Soviet science required honest appraisal of new ideas according to their own norms, even when outside authorities attempted to impose their views on the resolution of scientific conflicts.

Physics under Nazi Rule

Under National Socialist rule, German physicists, too, fell prey to extrascientific forces that influenced the content and direction of their work, and often their career paths. The majority of German physicists, like biologists, welcomed Nazi rule. They tended to be conservative individuals who, in spite of the large numbers of Jews among them, continued to embrace the strong anti-Semitic, antidemocratic, imperialistic, and nationalistic currents that dated in German science to the Wilhelmian empire.

Chauvinistic attitudes toward science were standard fare in Germany. At the outbreak of World War I, leading German scientists had enthusiastically banded together behind the kaiser in what they declared was a defensive war against an evil enemy, signing an "Appeal to the Cultured Peoples of the World." This proclamation revealed that prominent German artists and scientists were solidly behind their belicose, military-dominated government in what they claimed was an effort to protect the cultural heritage of Europe (Heilbron 1986, 70–71). A few scientists in private, and Einstein in public, rejected the proclamation. But it is no surprise that most scientists never trusted the Weimar leadership and welcomed the National Socialists to power.

The attack on the physics profession that followed the promulgation of the Nazi race laws paralleled the class-based assault on physics under Stalin. The intellectual migration that resulted devastated German physics, with perhaps one-quarter of German physicists forced from their jobs, mostly by the laws excluding Jews from the civil service beginning in 1933. Just as in the Soviet Union, most professors were government employees, so that it was relatively easy in both countries to impose state control even before the regimes had become totalitarian. As in Russia, so in Germany Jews had found greater opportunity for advancement in physics in the late nineteenth and early twentieth centuries; this partly explains the large number of Jewish physicists. The Jewish losses in physics were therefore proportionally greater than in biology.

TABLE 3.1   Nobel Laureates Who Left the Nazi Regime, 1933–1945

Physicists

Albert Einstein	James Franck
Gustav Hertz	Erwin Schrödinger
Viktor Hess	Otto Stern
Felix Bloch	Max Born
Hans Bethe	Dennis Gabor

Chemists

Fritz Haber	Peter Debye
George de Hevesy	Gerhard Herzberg

Medical Doctors

Otto Meyerhof	Otto Loewi
Boris Chain	Hans Krebs
Max Delbrück	

Source: Alan D. Beyerchen, *Scientists under Hitler: Politics and the Physics Community in the Third Reich* (New Haven and London: Yale University Press, 1977), 48.

Universities lost 17 percent of their teachers—Jews, liberals, and Marxists; institutes of technology, 11 percent. Attrition continued unabated. Beyerchen estimates that at least 25 percent of the physicists with positions in German higher educational institutions in 1932–33 were displaced during the Nazi period (Beyerchen 1977, 43–47).

Individuals motivated by professional jealousies were able to take advantage of a changed political climate to advance their careers. Theoretical physics lost its privileged position. The international aspect of physics came under attack. In this environment, even Werner Heisenberg, plainly a German patriot, was attacked for his support of modern physics, while Aryan physics was touted by its supporters as the only true German physics. Such non-Jewish physicists as Hans Euler and Bernhard Kockel were denied positions because of their leftist politics. Once physicists lost internal control over dismissals, they fought to protect the autonomy of their discipline in matters of appointments, curriculum, publications, and professional organizations. But in each of these areas, the promotion of Aryan physics under the auspices of the Reich Education Ministry politicized the process.

From the point of view of quality, the losses were equally disturbing. Twenty Nobel Prize winners were driven from their posts, including ten physicists (see table 3.1).

The effort to create an Aryan science freed from Jewish influence was especially prominent from 1933 until 1939. The leading physicists who remained to serve their nation—including Nobel Prize winners Max Planck (1918), Werner Heisenberg (1932), Philipp Lenard (1905), and Johannes Stark (1919)—were left to sort out what role modern theories, created in part by Jews such as Einstein, should play under a totalitarian regime founded on principles of racial purity. Lenard and Stark rejected these theories.

Stark's support of Aryan physics was both intellectually and personally motivated. He and Lenard were experimentalists who had made their careers early in the century by elaborating the Newtonian worldview. Physics to this time worldwide was dominated by experiment; as theory became useful, it opened a niche in academia for bright outsiders. Hence Jews tended to move into these career paths as universities opened positions in theoretical physics. Stark and Lenard resented what they believed was the diminution of the importance of experiment. More recent notions of light quanta, relativity, and quantum mechanics were anathema to them, as was the preeminence of theoretical physics generally. Stark often engaged in peevish priority disputes with other scientists, including Einstein. His hatred of relativity and Einstein had many roots. World War I "intensified Stark's nationalism and caused him to view Einstein's open pacifism and internationalism with intense disfavor." He was a chauvinist with respect to the roots of scientific achievements. Stark turned increasingly to technical physics and grew increasingly isolated at home and abroad (Beyerchen 1977, 103–11).

For Stark, that he had never been able to leave his chair at Würzburg University for a more prestigious one was evidence of a conspiracy. He accused theoreticians of sabotaging his efforts to move. But he had damaged himself professionally, using his Nobel Prize money against the spirit of the Nobel award to buy a porcelain factory and set up a private research laboratory. This led to his being ostracized by other German physicists. He was forced to resign his Würzburg post in 1922. Then, although a logical choice for president of the Institute of Physics and Technology in Berlin, Stark was passed over in favor of physical chemist Walther Nernst. Over the remainder of the decade Stark was considered and rejected six times for academic appointments. He responded to all of these personal disappointments by writing *Die gegenwärtige Krisis in der deutschen Physik* (The present crisis in German physics), which attacked relativity and quantum theory (Cassidy 1992, 342–43; Beyerchen 1977, 111–15).

Lenard was also a good candidate to embrace Nazism. He, too, rejected theoretical physics. His plodding, conservative approach was more conducive to work in experimental physics, not the rapidly unfolding areas of theory. Lenard carried deep-seated hostility toward Jews, blaming

Germany's defeat in World War I in part on them, and hated the Weimar Republic. He resented Einstein and the acclaim accorded relativity. He also despised Einstein's internationalist stance. But he reserved most of his hatred for the "abolition of the ether" that accompanied relativity theory. Like older physicists in many countries, he felt it was outrageous to replace the comfortably mechanical concept of a fluid bearing light waves with nothing more substantial than a set of equations (Lenard 1920).

During the 1920s, Lenard showed less and less academic restraint, affixing the label of "Jewish" to concepts with which he did not agree in the fight to save the ether. Lenard's anti-Semitism was inflamed after he refused to heed a Weimar decree to close his Heidelberg laboratory and lower the flag of his institute to mourn the assassination of Walther Rathenau, the liberal Jewish minister of the republic. Leftist workers and students then dragged Lenard from his lab, with some clamoring to throw him in a nearby river. Lenard and Stark began publishing a large number of anti-Semitic speeches to tout Aryan physics. Lenard turned toward consideration of the role of racial heritage in science. True science was experimental, national, and racially pure, he determined. The culmination of his contemplations was the four-volume *Deutsche Physik* (1936–37), in which *völkisch* concepts were front and center (Cassidy 1992, 342–43; Beyerchen 1977, 79–95).

Lenard claimed in *Deutsche Physik* that physics was not international, because everything that mankind produces "is conditioned by race, and by blood" (Cassidy 1992, 343–44). Another work characteristic of racist physics was the four-volume *Die Deutsche Physik* (1935), in which L. W. Helwig referred to the need to cleanse physics "of the out-growths which the by now well-known findings of race research have shown to be exclusive products of the Jewish mind and which the German Volk shun as racially incompatible with itself" (Speer 1970, 288). These writings were geared to free physics from "Jewish Marxist domination" and to fight against the "Jewification of German science."

Like its counterpart in the Soviet Union, Aryan physics rejected modern theoretical physics, especially quantum mechanics. Aryan physicists, like Marxists in the USSR, rejected the new physics on epistemological grounds. Like Timiriazev in Moscow, they were conservative experimentalists who intended to use the propitious political conditions to secure the professional recognition that they believed had been unjustly denied them. They channeled their anger into the anti-Semitic and antidemocratic sentiments prominent in National Socialism.

Inasmuch as it was a political more than a scientific movement, Aryan physics did not describe a standard approach to physical laws of nature.

Still, there were central features of Aryan physics (Beyerchen 1977, 123–40). It was anthropologically and racially based, with all leading concepts, its adherents unflinchingly claimed, originating among Aryan-German contributors. Experiment and observation were considered the only true bases of knowledge. Aryan physics was experimental, close to the facts of nature, and therefore highly useful for technology and industry, the better to promote economic self-sufficiency. Aryan physics rigidly embraced classical Newtonian physics—a paradoxical approach since its supporters also rejected mechanistic materialism as the foundation of equally vile Marxism.

The *völkisch* nature of this physics stemmed from the belief that the Nordic race had created not only mechanics but all experimental science. Nordic researchers had a penchant for observation, repetition, modesty, "joy in struggling with the object—joy in the hunt." The Jew, in contrast, had a predilection for theory and abstraction. Jews were behind the effort to abolish the ether conception. Science was never value-free or international. Rather, race and culture determined a researcher's perspective. International science was Jewish science and a threat to Aryan science.

Aryan physics had little appeal among most physicists because it rejected relativity and quantum mechanics, but it captured Nazi attention. Soon after Hitler came to power, Stark and Lenard tried to capitalize on their long-term personal support of the Nazis. They had been among the few leading scientists who supported Hitler during his brief imprisonment in 1924 for his role in the Munich Beer Hall Putsch in November 1923. Stark praised Hitler in print before the latter's rise to power, referring to his anti-Semitic ideas and his autobiographical tract *Mein Kampf* as evidence that Hitler was not a demogogue but a "great thinker" (Stark 1932). In March 1933, Lenard wrote directly to Hitler offering his services in personnel decisions affecting physics. Stark urged his fellow German Nobel laureates to join in a public declaration of support for Hitler in preparation for an August 1934 plebiscite. Heisenberg, Max von Laue, Planck, and the experimentalist Walter Nernst all refused, using the argument that politics and science should not mix. Stark responded that support of Hitler in the plebiscite was not a political act but an "avowal of the German Volk to its Führer." Stark added that the other physicists were being political by honoring Einstein but not speaking in favor of Hitler. Hitler could hardly forget the support two Nobel Prize winners had given him when he assumed control of the total state (Walker 1989a, 64; Walker 1989b, 61–62).

Heisenberg tried to avoid recognition of this distasteful display of racism. He promoted theoretical physics as being in step with experimental studies and based on "cogent empirical methods." Theoretical physics was getting less support in the first years of Nazi rule, so Heisenberg defended it for its role in technological progress. He continued, perhaps dangerously, to mention Jewish scientists in the same breath as representatives of Nazi physics like Lenard. His public declarations led some Nazis to call for him to be interned in a concentration camp. While the leading Nazi ideologue Alfred Rosenberg agreed with this sentiment, he could not follow through on internment, owing to Heisenberg's international stature (Cassidy 1992, 335–37, 340–42). But Stark was able to turn his Nazi credentials into official positions that gave him immense influence in science policy, positions where he could work to derail Heisenberg's career.

Simultaneously, the Reich Education Ministry under Bernhard Rust supported Nazi academic organizations such as the Teachers League and the Students League. This meant a period of flowering for Aryan physics, yet set Stark on a collision course with Rust, who, like Stark, intended to dominate Nazi academic policy for personal gain. Stark was made head of the prestigious Physical Technical Reich Institute (Pfetsch 1970). Stark tried to reorganize all scientific research and publications under the presidency of the institute. He submitted his own candidacy for presidency and for membership in the Prussian Academy of Sciences as Einstein's successor after Einstein resigned. Many in the physics profession either remained aloof or fully supported the new regime, but others hoped that back-room diplomacy would keep Stark from realizing his goal of domination. The intervention of Laue, who was president of the German Physical Society, indeed thwarted Stark's plans. Then Rust appointed Stark president of the Notgemeinschaft der deutschen Wissenschaft. This appointment gave Stark authority over significant funds but proved his undoing when he tried to use it to control all science, putting him in conflict with others who wished to do the same thing. He alienated all the Nazi Party insiders in the Education Ministry (where many members of the fanatically Nazi SS [*Schutzstaffel* or Blackshirts, Hitler's most loyal troops, under the command of Heinrich Himmler] were to be found) and elsewhere (Beyerchen 1977, 115–22; Cassidy 1992, 344–45). Stark had lost power by 1937, but not before subjecting Heisenberg to blistering personal attacks.

During the early years of the Third Reich, Lenard, Stark, and their associates blocked the attempt to appoint Heisenberg as successor to his Munich university teacher Arnold Sommerfeld, a well-merited appointment already approved by the university and the Bavarian Ministry of Culture. The attacks on Heisenberg stirred the German physics community to retaliate, prompting the government to reappraise modern theoretical physics (Walker 1989a, 63–64).

The high point of the Aryan physics movement came in 1936, when Stark, Lenard, and their supporters attacked mainstream German physicists. They had gathered around them a group of young scientist-followers. They attacked Heisenberg in the *Völkische Beobachter*, the semi official paper of the party, and in speeches. Then the SS newspaper, *Das Schwarze Korps*, attacked Heisenberg and other German physicists as "white Jews," that is, as individuals of Aryan heritage but under the influence of Judaism, in this case Einstein's physics. The attacks led a significant number of students to decline to study with Heisenberg and other theoreticians. If such charges continued and strengthened, not only Heisenberg's position but his freedom could be at risk (Walker 1989b, 61–62).

In defense against charges that he was a "white Jew," and in pursuit of what was considered rightfully his Munich professorship, Heisenberg took advantage of personal contacts and the absence of a monolithic Nazi science policy. Granted, he had to tread carefully, and the effort took time and involved personal hazard. Heisenberg first wrote to the Reich's minister of education to defend his honor and to demand that personal attacks such as those in *Das Schwarze Korps* be prohibited, threatening to leave Germany should this option not be open. He then wrote to Reichsführer-SS Heinrich Himmler. If Himmler approved of Stark's attack, Heisenberg would resign; if he disapproved, Heisenberg requested restitution of his honor and protection against further attacks. Heisenberg's grandfather belonged to a hiking club that included Himmler's father, through which Heisenberg's mother grew acquainted with Mrs. Himmler, who delivered the letter to her son. The letter led Himmler to pursue the matter in depth (Cassidy 1992, 384–87; Walker 1989b, 62).

Still, Heisenberg had a number of disconcerting experiences. First, the Nazis were strong in Bavaria and were able to promote opposition to him there among student and professorial groups. Nazi students used their journals, *Deutsche Mathematik* (German mathematics) and *Zeitschrift für die Gesamte Naturwissenschaft* (Journal for the entirety of science) throughout 1937 to produce "a rising torrent of anti-Semitic Nazi science propaganda." This torrent targeted Heisenberg, accusing him also of being politically unreliable and a homosexual. These charges delayed once again Heisenberg's appointment to Sommerfeld's chair, even though the candidates proposed by Nazi teaching organizations as alternatives paled in comparison. Ludwig Bieberbach, acting vice-chancellor of the University of Berlin and dean of his faculty, contributed a supposed proof that achievement in mathematics was linked to race (Mehrtens 1994; Cassidy 1992, 388–93; Walker 1989a, 66).

Finally, Heisenberg was interrogated at SS headquarters in Berlin. Fortunately, one of the SS investigators had earned a doctorate under Laue at the University of Berlin and was inclined to understand Heisenberg's position. The SS investigation revealed no political or Semitic transgressions, merely "a harmless apolitical academic." The personal intervention with Himmler of Ludwig Prandtl, Göttingen professor of applied mathematics, ultimately got Himmler to act. In a discussion with Himmler, Prandtl gave lip service to the charge that non-Aryan physicists had tainted theoretical physics but pointed out that some non-Aryan scientists did first-rate work; in other words, he drew a distinction between the person and the physicist. Himmler told Heisenberg in summer 1938 that he disapproved of the personal attacks against him, had proscribed them, and asked that Heisenberg merely separate the man, the Jewish Einstein, from his research, the modern physics (Cassidy 1992, 388–93; Walker 1989a, 66).

In this way, theoretical physics remained largely intact, although Jewish names, like those of forbidden colleagues and political outsiders in the Soviet Union, could not be mentioned in a positive vein. The prohibition was official; in private discussions the physicists continued to use the names of Einstein, Born, and Franck. But Stark and Lenard's Aryan physics did not simply disappear. Heisenberg had more battles to fight after Himmler's intercession. In the same way that cultural revolution in the USSR had changed the traditional way of entering university, defending one's degree, and making appointments, so in Nazi Germany the party apparatus had gained significant influence over all these areas of professional concern and autonomy. Rudolph Hess, deputy leader of the Nazi Party, encouraged party academic organizations—the Students and Teachers leagues—to make independent recommendations about appointments to the Reich Education Ministry, which had the final say. Arnold Sommerfeld officially announced his long-expected retirement in 1938, hoping this would finally push the ministry to appoint Heisenberg his successor to the theoretical physics chair. The Munich University faculty produced two lists, one with Heisenberg at the top, the other with the Teachers League's recommendations of three pedestrian applied physicists. The Teachers League endorsed Aryan physics by urging "a return to the natural and the pictorial." It nominated men who rejected the abstract mathematical formalism of Einstein's relativity theory for a close reliance on experimental data. In the meantime, Sommerfeld's chair remained unfilled.

Heisenberg spent much of 1937 and 1938 at home and abroad countering the latest opposition. He gave lectures in subjects not entirely based on theoretical approaches, for example, cosmic-ray physics. In the winter of 1938–39, Heisenberg took his theoretical work in nuclear physics on the lecture circuit again, claiming that his explanation of the lifetime of the meson particle was "a striking confirmation" of the relativity principle, although he did not mention Einstein by name.

Aryan policies and intrigues led to the international isolation of Nazi science. Whether autarkic relations were desired by the government is not the point. German scientists grew isolated from the mainstream of international science, particularly through a decline in the number of foreigners visiting Germany, holding memberships in learned societies, and subscribing and contributing to German scientific journals. There were also restrictions on Germans who traveled abroad (Beyerchen 1977, 71–76). Heisenberg was caught up in this isolation, largely because of his own doing. During a trip to the United States in summer 1939, Heisenberg met criticism. It seemed to American colleagues that his treatment and that of other physicists at the hands of the Nazis was reason enough to emigrate. And then there was Kristallnacht, the night of November 9–10, 1938, when in a national frenzy of anti-Semitism whipped up by Nazi hooligans the stores and homes of Jewish shopkeepers were looted and their contents smashed on the streets. There could be no doubt about the attitude of the regime toward Jews. Yet Heisenberg and other mainstream physicists seem to have ignored Kristallnacht. They wanted to remain apolitical in their relation to the regime. Heisenberg had chosen to stay and maintain his allegiance to the Nazi government because he was a German patriot. He hoped the regime would change its tune, moderating over time, if not be replaced. And if there was to be war, he wanted to help his country, no matter the politics of the leadership. He also felt a loyalty to the German physics community. According to historian David Cassidy, Heisenberg had come to identify his activity in Germany as central to the survival of modern physics there (Cassidy 1992, 397–414).

Surely it was difficult for patriotic scientists to choose the right course in response to Nazism. Many honest Germans believed that by using quiet diplomacy rather than visible public protest they might moderate Hitler's behavior. Others assumed the Nazis were a short-lived regime, not representing the true spirit of Germany. And many, no matter what else they believed, welcomed National Socialism's "call to national cultural renewal, unity, and glory." In an intellectual biography, John Heilbron indicates the difficult position of leading scientists such as Max Planck when faced with Hitler's rise. Planck felt compelled to accommodate the Nazi regime to protect science from the interference of lesser men. He urged his colleagues not to declare in favor of their Jewish colleagues, since this would cause many others to speak out against them in order to take their positions. Early in the Nazi era, Planck urged Heisenberg not to resign but to remain as a guiding light to the young on appropriate behavior. He avoided public declarations against race in science and totalitarianism. In an interview with Hitler, Planck convinced himself that the dictator had nothing against Jews. Planck asked Einstein to consider not the motives behind his decision in March 1933 not to return to Germany but the consequences of making it public. Planck felt certain that the consequences of Einstein's actions for Germany's Jews and for the Prussian Academy of Sciences would be deleterious. The raucous condemnation of Einstein that followed, orchestrated by Nazi education minister Bernhard Rust, showed just what a bind well-intended individuals could be in when faced with an irrational regime.

Planck then had to face the wrath of Stark and Lenard when he proposed Max von Laue as Einstein's successor in the academy. The Aryan physicists saw Laue, Einstein's close friend, as little better an example for German youth. Planck was central to efforts to defeat Stark's candidacy for the academy, which also retained its Jewish members, other than Einstein, and its Jewish employees. And he also protected the Kaiser Wilhelm Society from Aryan physicists, even securing financing from the Rockefeller Foundation in 1934 to establish a physics institute under the auspices of the society. But Planck was finally forced to retire from his positions of power and responsibility and ultimately could not defend science from extremely damaging Aryan interference. He, like most of his fellow physicists remaining in Nazi Germany, failed to comprehend that scientific activity under Hitler was inherently political (Heilbron 1986, 149–203).

It was not so much the defense of modern physics by such individuals as Heisenberg as the difficult conduct of the war that convinced government, military, and party officials alike that continued training of young specialists and production of armaments were more crucial to the Reich than any concern about possible harm to German physics by Jewish influence. They now sought rapprochement with the physics community, abandoning vocal support of Aryan physics. Wolfgang Finkelnburg, a member of the Teachers League, arranged a conference in Munich in 1940 to debate the new physics at which members of the Aryan physics movement were forced to concede that theoretical physics was a necessary part of physics, that further investigation into relativity was scientifically justified, that relativity theory did not signify philosophical relativism, and that quantum and wave mechanics were essential to understanding atomic processes.

Carl Ramsauer, a leading physicist at German General Electric, newly chosen head of the German Physics Society, and a former student of Lenard, contributed to the defeat of Aryan physics. He compiled statistical evidence on prizes, scientific citations, and publications that showed how American physics now outstripped German in university training and had moved to the forefront of theoretical physics, while German physics lagged in research and education financing. He hinted strongly that political interference into professional appointments and standards was the culprit here. And he argued that only the new physics could provide the new weapons needed to wage war successfully. Finally, the government and party resumed granting significant appointments in universities in theoretical physics to individuals opposed to Lenard and Stark. Heisenberg got a professorship at the University of Berlin along with directorship of the Kaiser Wilhelm Institute for Physics, but only later the chair of theoretical physics in Munich (Walker 1989a, 69–72, 79–85).

Still, just as in the USSR, where the Party's ideological scrutiny required careful rendering of physical concepts or reference to individuals considered anathema to the state, so in Nazi Germany the state was always capable of using its power in an arbitrary and capricious fashion against scientists. Even after the decline of the Aryan physics movement the publication of texts might be held up if they contained too many references to Einstein the man. And only in 1943 was Heisenberg able to publish an authoritative article to rebut Aryan physics once and for all, entitled "The Value of 'Modern Theoretical Physics.'" As Walker points out: "Both the state and party realized that apolitical scientists were not merely acceptable; in the absence of a sufficient number of researchers who were both capable and absolutely loyal to National Socialism, apolitical science was a necessity." (As we have seen earlier, the same point holds for the USSR in the 1920s and 1930s.) The Nazi Party, for its part, came to believe that the dispute between representatives of the new physics and of Aryan physics was a professional, intramural dispute, not a political one, with both groups loyal to the regime (Walker 1989a, 71–75, 84).

The Totalitarian Bomb

World War II changed forever the nature of warfare: civilian targets assumed strategic importance; weapons enabled the attacker to be far removed from the victim; and levels of casualties in single attacks reached the tens of thousands at Dresden, Tokyo, Hiroshima, and Nagasaki. When it came to the atomic bomb, the core differences in the nature of scientific activity in Nazi Germany, the Soviet Union, and democratic countries like the United States were often blurred. In each country, patriotic scientists sought to help their government develop nuclear weapons in short order, before sworn enemies did. In each, few individuals questioned whether their participation was morally right. In each, the requirements of secrecy prevented wide-ranging discussion of the moral questions surrounding the use of weapons of mass destruction. The absence of questioning may have been the result of the belief that research was concerned with questions of fact (Could a bomb be built?), not questions of moral issue (Ought we to use such horrible weapons?). (For issues of responsibility of scientists involved in the development of weapons of mass destruction, see Badash 1995.)

In spite of the similarities that characterize the atomic bomb projects in these three countries, there were important differences. First of all, a comparison of the Nazi and Soviet atomic bomb projects reveals in high relief the strengths and weaknesses of the organization and administration of science under totalitarian regimes, putting to rest the claim that totalitarian regimes are always more efficient than pluralist forms of government at marshaling resources and personnel to achieve desired ends.

The USSR was able to make personnel, institutes, and materials a priority for its bomb, diverting resources from housing, environment, and medicine for the duration of the Cold War. Yet even when questions of power and authority are resolved ruthlessly through single authorities, a project will stumble if it fails to get their attention, or if the wrong path is chosen in the absence of free debate over alternatives. (The problem is most serious with "big" science; smaller projects often find the wherewithal to get started under the umbrella of other projects.) Further, the centralized decision making and distribution of resources may exist only on the surface, obscuring political and scientific rivalries, and then efforts to distribute resources optimally will be handicapped. This was the case with the bomb project in Germany.

In both regimes, however, the importance of the atomic bomb project effectively ended ideologists' efforts to condemn the new physics. In the USSR the antirelativists were silenced when Stalin learned that the new physics could contribute to the state's military might. Once this happened, Stalin's atomic bomb moved ahead rapidly. Scientists achieved success with the help of espionage, to be sure. But a combination of other historical and political factors was far more important in securing the rapid construction of nuclear weapons in the USSR. The first

Soviet explosion was in August 1949, many years earlier than most Western observers had anticipated. They need not have been surprised. Soviet physicists were among world leaders in nuclear physics, having followed closely and participated in the almost miraculous unfolding of knowledge about the nucleus during the 1930s. At a handful of institutes in Leningrad, Moscow, and Kharkov, physicists built cyclotrons, Van de Graaff accelerators, and other devices to peer into the nucleus. They had only grudging government support, since science policy makers could not see any immediate industrial applications from their research. But on the eve of World War II, Soviet physicists were prepared to engage in nuclear arms research (Josephson 1987).

In each country, scientists convinced their country's leaders to join a race for nuclear weapons, having alerted them about the contest in the first place. In the United States, in 1939 Leo Szilard talked Albert Einstein into signing his name to a letter to President Franklin Roosevelt in which the specter of a Nazi bomb was raised. After other American and British scientists added their warnings, the Manhattan Project commenced in 1942 (Badash 1995). In the USSR, Georgii Flerov played the roles simultaneously of Szilard and Einstein, writing to Stalin himself in 1942 to call for a similar project. When espionage showed that the Americans were up to something, Flerov's requests were granted, with Igor Kurchatov appointed the director of the project in "laboratory no. 2" on the outskirts of Moscow in 1943. Kurchatov was given great authority to requisition personnel, instruments, and supplies.

Until the Americans dropped bombs on Hiroshima and Nagasaki in August 1945, Kurchatov's project moved ahead slowly. Then Stalin instructed secret police chief Lavrenti Beria to bar no expenses, and a broad program commenced that involved building an experimental reactor to understand better the properties of fissile isotopes and neutrons, followed by several industrial reactors for producing plutonium for the bombs, isotope separation plants, and a site for testing. The nuclear industry expanded under Kurchatov's able direction and the close supervision of the project by Stalin and Beria at every step of the way.

Espionage, in particular the efforts of physicist Klaus Fuchs to funnel secrets from the American project to Soviet intelligence, played a crucial role in the Soviet success: not so much by providing detailed blueprints (which shortened the Soviets' efforts by perhaps a year or so) but by convincing Beria and Stalin that Kurchatov and his colleagues were on the right path, so that they were inclined to give the physicists greater independence. Soviet physicists themselves acknowledged this aspect of Fuchs's contribution. The Soviets also benefited from knowing that the Americans had succeeded and from access to the widely published, unclassified Smyth report on the Manhattan Project, which was quickly translated into Russian and distributed throughout the scientific community in the USSR (Smyth 1946). But Kurchatov still had to build reactors, run experiments, and so on. The Soviet hydrogen bomb was developed entirely indigenously based on ideas from Andrei Sakharov, Vitali Ginzburg, Evgenii Tamm, and others. In a word, the USSR had the native scientific talent and political will to become a nuclear power within four years after the United States had detonated its first atomic bomb and to develop hydrogen bombs at roughly the same time as the Americans (Holloway 1994; Rhodes 1995).

Scientific success and an end to antirelativist pressures were one thing. Stalin was still unwilling to permit scientists to discuss the geopolitical and moral ramifications of nuclear weapons. Soviet scientists could not protest against nuclear weapons programs without bringing down the wrath of their government. Nor could Soviet scientific organizations dedicated to arms control be of more than limited effectiveness. The Communist Party designated representatives to organizations, such as Pugwash, that were intended to build confidence among American and Soviet representatives interested in arms control. Some of these individuals were independent-minded, but many only served intelligence purposes for the government. Rare, therefore, were scientists like Andrei Sakharov who spoke out against the arms race.

Sakharov, father of the Soviet hydrogen bomb and later a human rights activist, became one of the leading spokespersons in the arms control effort. He was a product of both the Stalin era and the so-called de-Stalinization thaw initiated by Nikita Khrushchev. As a patriotic physicist, he willingly devoted his talents to the development of the hydrogen bomb. Yet, in response to the twentieth Party Congress at which Khrushchev had attacked the human costs of Stalin's rule, Sakharov began to explore his involvement with the military-industrial complex. Recognizing the cynicism of the authorities toward scientists, he decided to enter the political sphere directly. In the late 1950s, he promoted a unilateral test ban. When he realized that continued nuclear testing was out of the hands of scientists, controlled by short-sighted and aggressive military men, and had significant nonthreshold biological effects, his decision to cease participation in the arms race became firm. Sakharov argued that peaceful coexistence with the West and a ban on nuclear weapons testing were the only way to avoid the destruction of civilization. He then joined other physicists in the fight against Lysenkoism.

There were, of course, limits on physicists' authority in the Soviet Union. Sakharov overestimated the reach of the de-Stalinization thaw and underestimated the power of the military, who wished to expand the USSR's nuclear arsenal. Khrushchev tired of Sakharov's meddling in what he viewed as purely political, not scientific, affairs, so that Sakharov's voice became but one cry in the wilderness. Leonid Brezhnev, who succeeded Khrushchev, could not tolerate Sakharov's increasing activity in the struggle for human rights in the Soviet Union and ordered him exiled to the city of Gorky where he remained under constant surveillance (Sakharov 1990, 96–97, 197–204, 207–9, 215–18, 267–68; Holloway 1994).

The openness of pluralist regimes naturally encourages scientists and citizens to address the morality of nuclear weapons. And this fact, paradoxically, made the lives of American scientists more difficult, since their public existence forced them to acknowledge their participation in the production of weapons of mass destruction used at Hiroshima and Nagasaki (Sherwin 1975), whereas Soviet physicists were treated in their country as unquestioned heroes. Scientists in totalitarian regimes also can take a moral stance, but the greater risk of rejection by society restricts this action to the most independent souls.

Hitler's Atomic Bomb

German physicists, too, only belatedly discussed the morality of their efforts. Under Nazi rule, patriotic scientists such as Heisenberg worked to develop nuclear weapons. These scientists became convinced early on that their efforts to create atomic weapons would not reach fruition during the war (Walker 1989b, 223). After Germany's defeat, some used their slow progress to claim later on that they had been morally and ethically opposed to producing such a horrible weapon for the immoral Nazi regime. But the evidence reveals that German scientists cannot take the moral high ground. They turned a blind eye to the moral standing of the regime. Their research was geared toward military applications of nuclear physics. They knew German leaders would use any weapon they developed in battle against England, the USSR, or the United States.

The Nazi regime needed mainstream physicists for its war effort and had to make ideological concessions to them and against the Aryan physicists. The ideological interference was paradoxical, since scientists largely saw the Nazi regime as more compatible with science than the Weimar Republic because of their traditional trust in a nationalistic, totalitarian state. Such physicists as Otto Hahn, Max Planck, and

Werner Heisenberg did not resist the regime but accommodated to it in many cases, even when it passed the race laws so damaging to their Jewish colleagues (Walker 1989b, 229–33). Heisenberg and his colleagues directly and indirectly used their efforts in the atomic bomb project and other research efforts of military significance to counter the Aryan physics movement.

Heisenberg claimed in 1947 that political interference and war conditions handicapped physicists' efforts, that building a bomb was too big a job for Germany in wartime, and that the physicists wanted to keep control of the project and focus on a reactor; hence they were spared the decision about whether to build a bomb. Samuel Goudsmit, a physicist who led the secret ALSOS mission to Germany one step behind the advancing troops to evaluate the achievements of the Nazi bomb project and capture its leading scientists for the United States, says the claim of German physicists that they had decided not to build a bomb was a cover for their immorality and for the fact that they had tried to build a bomb but didn't know how (Goudsmit 1947; Powers 1993, 430–35).

Many of the German physicists involved in the Nazi bomb project— Otto Hahn, Walter Gerlach, Max von Laue, Werner Heisenberg, and others—were interned at Farm Hall in England for the last half of 1945, where their conversations were secretly monitored and recorded. From conversations on the evening of August 6, 1945, when the Germans were informed of Hiroshima, it is clear that Heisenberg grossly overestimated the quantity of the isotope uranium-235 necessary for an explosion and had not figured out how a self-sustaining chain reaction would occur in a bomb. In an informal colloquium that Heisenberg delivered to fellow internees at Farm Hall on August 14, he had more of the science worked out but failed on some issues (Bernstein 1993; Powers 1993, 435–52).

More revealing of Heisenberg's avoidance of moral issues was his service for the Nazi regime as a cultural ambassador. We recall Heisenberg's summer 1939 lecture tour of the United States and his silence over Kristallnacht. During the war, Heisenberg willingly traveled to conquered lands as the regime's representative to display the glorious scientific achievements of German civilization. He was devoted to his country, his political troubles with Stark, Lenard, and their followers notwithstanding. Heisenberg justified his return to Germany to work on nuclear fission by saying, "Because of their sense of decency most of the leading scientists disliked the totalitarian system. Yet as patriots who loved their country they could not refuse to work for the Government when called upon. . . . Fortunately, they never had to make a moral decision, and this for the reason that they and the Army agreed on the utter impossibility of producing a bomb during the war" (Powers 1993, 456–57).

In fact, administrative rather than moral factors interfered with the Nazi bomb project. Goudsmit wrote a book about the ALSOS mission to disprove the myth that Nazi military supremacy was based on science. The mission revealed that organizational, scientific, and financial missteps doomed the Nazi bomb from the start. Goudsmit bluntly states, "Too many of us still assume that totalitarianism gets things done where democracy only fumbles along, and that certainly in those branches of science contributing directly to the war effort, the Nazis were able to cut all corners and proceed with ruthless and matchless efficiency. Nothing could be further from the truth." Goudsmit says the Nazis failed because they did not cut through ministerial red tape, complacently assumed their science could secure success before any competitors, and allowed politics blindly to interfere in the affairs of science (Goudsmit 1947, xi–xii).

In Germany, in spite of the leader principle, nuclear research was never centrally coordinated. Groups of scientists and administrators competed for influence over the project. On one side stood the Kaiser Wilhelm Society, Heisenberg, and the Ministry of Armaments and Munitions. On the other side stood the Reich Research Council under the jurisdiction of the Education Ministry, the stronghold of Aryan physics, with control over equipment and research materials.

Heisenberg was bitter about the Reich Research Council's neglect of nuclear research, the lack of funding and materials, and the drafting of capable scientists into active service. He worried about America's head start in research. But Heisenberg was rarely encouraging to potential sources of funding about whether nuclear bombs might be developed for use in the current war, although he acknowledged the theoretical possibility. When the minister of armaments and munitions, Albert Speer, offered his ministry's resources to build a huge cyclotron, as large as or larger than any in the United States, Heisenberg declined the offer, focusing instead on the importance of starting small and building on experience (Speer 1970, 224–27).

Speer attempted to put the bomb project on the front burner in 1942. He was intrigued by superweapons and, as minister, had access to manpower, equipment, and materials. But Army Ordnance had already decided not to shift nuclear power research up to the industrial level of production. Shortages of the heavy water necessary for a reactor fueled with natural uranium, delays in development of isotope separation, and a series of other technical problems also contributed to the lag in atomic research specifically for a bomb. Further, Hitler and Hermann Göring, Hitler's second in command and head of the war economy, never developed an abiding interest in the nuclear project. Göring's lack of interest was not surprising, given the immediate demands of the war effort and his confused understanding of science: he once suggested that Speer make locomotives out of concrete, since there was not enough steel available. Hitler, for his part, seemed not to have grasped the revolutionary nature of a nuclear weapon and saw no short-term reason to pursue it (Speer 1970, 224–26; Walker 1989b, 78–90, 204–10). Speer gives ideology as one of the reasons for failure to pursue atomic possibilities, for Lenard gave Hitler to believe that the Jews were exerting a seditious influence on nuclear physics and relativity theory. Indeed, Hitler occasionally referred to nuclear physics as "Jewish physics" and cited Lenard as his authority (Speer 1970, 228).

Ideology, Isolation, and Einstein

Americans tend to believe that German and Soviet physicists involved in the development of nuclear weapons had a lower moral standing than American scientists connected with the Manhattan Project. In totalitarian regimes service to the nation dictates silence on issues of ethics and morals that are discussed openly in pluralist systems. The National Socialist and Soviet governments used draconian measures to prevent German and Soviet physicists from engaging in open opposition of any kind; few even thought to oppose weapons development. These measures, which extended well beyond the power of the purse to include firings, ideological intrusions into the sphere of scientific research, and executions, grew out of cultural revolution.

Hence cultural and administrative revolution that accompany ideologization distinguish science in totalitarian regimes from that in pluralist regimes. Ideologization affects norms of scientific behavior. Ideology specifies internal and external enemies who are to be fought in every sphere. To ensure success in this struggle, science is placed under administrative control. This leads to the replacement of more democratically run bureaucracies and institutions with top-heavy, centrally managed ones. The individuals who are employed in these organizations usually represent party organizations and strive to harness scientific research to the needs of the state through planning, financial, and supervisory means. Researchers are held accountable to the state. Stress is placed on applied research at the expense of basic science. Ideologization enables individuals who are professionally and personally motivated to step outside of normal channels, such as peer review, to damage the careers of individuals whose views they dispute.

Twentieth-century science is international. In totalitarian regimes, the state tends to impose autarkic relations. Yet autarkic trends already exist in the nationalistic philosophies of science that are advanced by supporters of the regime. They stress central planning and other forms of control of the research agenda to ensure that the domestic ideological agenda is paramount. In Nazi Germany, the Aryan physics movement rejected internationalism for a racist doctrine that saw all good physics as having *völkisch* roots. In the Soviet Union, as part of a class-based doctrine, Marxist ideologues rejected internationalism for "socialism in one country" and saw dangers to the proletariat lurking in the epistemological implications of the new physics. The alleged enemies were members of the international bourgeoisie and individuals connected to conspiracies against the Communist Party and Stalin. In Russia as in Germany, it did not help that Einstein himself was an internationalist. Paradoxically, under both regimes scientists were valuable not only for their technical prowess but for the prestige they generated abroad. In some cases, they served as cultural ambassadors of their outcast nations.

We began this chapter with a discussion of an apple falling to the ground in four different cities. Consider now Einstein's presence in each of those cities in 1935. In Cambridge and London, he is a hero whose opinions on matters of science and religion, philosophy and government are all respected. Under totalitarian regimes, Einstein is persona non grata. In Berlin, he is an enemy of the state as a Jew, a theoretician, an internationalist, and a pacifist. In Moscow, he is also an enemy as a Zionist and an alleged philosophical relativist and subjective idealist. Owing to the ideologization of science, physicists in totalitarian regimes had to employ the sort of physics to which Einstein had contributed with great care.

Note

1. The Deborinites authored three major treatises on relativity theory: V. P. Egorshin's *Estestvoznanie, filosofiia i marksizm* (Moscow, 1930), S. Iu. Semkovskii's *Teoriia otnositel'nosti i materializm* (Kharkov, 1924), and Boris Gessen's *Osnovnye idei teorii otnositel'nosti* (Moscow, 1928). The works share a common respect for relativity theory, a recognition that classical explanations for matter, electricity, and motion were inadequate in certain cases, and a belief that relativity theory and dialectical materialism were compatible.

Technology and Politics in
Totalitarian Regimes

TECHNOLOGIES ARE SYMBOLS of national achievement. They demonstrate the prowess of the nation's scientists and engineers. They are central to national security strategies. They serve foreign policy purposes through technology transfer. They entrance a public who can become intoxicated with the artifact's symbolism and overlook its potential dangers to society (and at other times provoke fear and dislike). We need only think of the space race between the United States and the USSR, or other technological posturing between the two superpowers, to comprehend the importance of technology in securing a regime's legitimacy at home and abroad.

Skyscrapers and apartment housing, subway systems, assembly lines, canals and bridges, hydropower stations, and nuclear reactors have an imposing physical presence. They also have what has been called "display value" (Smith 1983; Frost 1991; Hughes 1989b). Display value includes the social, cultural, and ideological significance of technology. While countries differ in terms of economic and political organization—market or centrally planned economy, single-party or multiparty system, centralized or decentralized decision-making apparatus—the display value of large-scale technologies applies to all.

Yet surely the place of technology in totalitarian regimes differs from that in pluralist regimes. If economic, political, and social concerns shape the practice of biology and physics in totalitarian regimes, then it should come as no surprise that technology, too, has a particular style in those regimes. At first glance technology would seem to be value-neutral, serving the rational ends of achieving a desired outcome in the "one best way." This way means efficiency maximization. Technologies are various devices, techniques, or systems intended to give us control over the natural environment—and also over our political, economic, and social structures. The latter include scientific management

75

for industry, the gathering, collating, and analysis of data for national planning, and so on. Engineers strive for efficiencies in production by optimizing the use of labor and capital inputs. They work toward these ends through planning to set prices, allocating raw materials and market share, and designing shop layouts and material flows (Ellul 1964).

The "one best way" distinction is crucial, for it implies that given any engineering problem the solution will be universal, based on engineering calculations that employ the scientific method. The "one best way" means that rockets and jets the world over resemble each other because other designs would not fly. All hydroelectric stations, subways, bridges, and skyscrapers share essential materials, structural elements, and components, or they would not stand. The first-glance differences between technologies in different settings reflect, literally, superficial elements: the skin of glass and steel or aluminum and plexiglas of a skyscraper, for example. You could go so far as to say that functional efficiency determines design. Yet technologies are more than components assembled in the "one best way" to create a large system. Economic and social obstacles as much as technical ones must be overcome to ensure successful diffusion of technology. Capital, political and human organizations are vital to technology (Hughes 1989a).

Engineers trained in a given milieu tend to accept the broader cultural values of their system. What are rational means for achieving desired ends in one society may be abhorrent in another. For example, the mass production of consumer goods through the "American system" of interchangeable parts and Fordism (standardized production; a controlled and steady flow of energy and materials in production processes from acquisition to the assembly line; and mass production to lower unit costs) will be crass materialism to conservative German engineers. The factory assembly line symbolized the exploitation of the proletariat to Soviet engineers. On the other hand, Soviet leaders idealized Taylorism (a doctrine of scientific management in industry) and established an officially sponsored materialism. And when the ends are full employment, social welfare, inexpensive housing, or universal health care (and not *simply* the design of a jet engine!), disagreements over the means and ends pour forth.

Take the example of public housing. Recognizing that their expertise could be used to achieve social goals, and responding to housing needs, engineers in Nazi Germany and the Stalinist USSR sought a prominent role in factory organization, housing, and urban planning. In the USSR, a Marxist urban industrial ideology held sway. Urban centers swelled in the 1920s and 1930s as peasants streamed into cities. Stalin encouraged this behavior by cutting capital investment in the countryside to focus on the creation of heavy industry and by forcing the peasantry into hated collected farms. Planners' preferences held sway in this centrally planned economy. Housing had to be built rapidly. Why not use inexpensive, standardized designs based on prefabricated forms that could be assembled rapidly by unskilled and illiterate workers into dwellings?

Soviet housing was proletarian in its minimal space, threadbare appointments, and shared bathrooms. This housing frequently incorporated the "collectivist" ethos in communal kitchens, child care facilities, and rooms for workers' clubs, but these were introduced more often to cut costs than to uphold a proletarian social ideal. The apartments and clubrooms, like the factory itself, also had a political function as the appropriate setting for the Communist Party to employ various media (radio, film, mass publications, meetings) to educate the masses about Stalin's programs. In Nazi Germany, these communal means and ends were rejected out of hand as anathema to the *völkisch* peasant and reviled as "Bolshevist." More appropriate for the German were thatched-roof cottages that showed his organic and blood ties to the soil.

The Totalitarian Machine

Does the machine, the symbol of the engineer, have the same effect on societies everywhere? Do all of the world's engineers employ machines for the same ends? Is the universal goal of the machine such efficiencies as increased output per unit of input, economies of scale through mass production, and speed of output? More to the point, can the engineer make rational, optimalistic choices in a totalitarian system? The answer is often yes, but the path to that answer has been arduous in every system.

Yet the ideological underpinnings of National Socialism and Soviet Marxism differed significantly. Nazism was an antiurban, racially based ideology. According to its myth, Aryan "settlers" whose blood rooted them organically in the German soil created a technology that served *völkisch* needs, not the profit motive of international capitalists. At the same time, German technological achievements—for example, in the chemical and automotive industries—were pioneering efforts that displayed complex elegance. So there was a disjunction between the advanced state of German technology and the Nazi myth, described by historian Jeffrey Herf as "reactionary modernism" (Herf 1984).

Soviet technologies were intended to reflect the collectivist ethos of serving the basic housing, transport, and food demands of the masses. Simultaneously, they served state goals of economic self-sufficiency and military might. The construction sites were also forums for educating the unskilled workers, not only about technical details but also in the messages of Stalinism. The result was bland, functional designs in which workers' safety and environmental concerns played a secondary role.

In spite of these differences, several features distinguish technologies in totalitarian regimes from those in pluralist regimes. The most obvious is the fact that the state serves as prime mover behind development and diffusion. Whether in Soviet research institutes or Nazi ordnance laboratories, this engenders "big science" approaches to research and development. The absence of market forces and the exclusion of the public from decisions about how or whether to diffuse a technology permits the development of technologies that persist no matter their questionable efficacy or environmental soundness. These characteristics also apply in the nonmarket sectors of democracies, notably defense industries, which are infamous for projects that waste billions of dollars.

A second feature of totalitarian technology is overly centralized administration of research and development. This is not surprising in countries of state socialism, such as the former USSR, where the state owns most of the means of production. But in fascist regimes, too, the persistence of private property is tempered by centrally funded projects that rely on the state for their impetus. Major industrialists prosper in close cooperation with the state, while smaller businesses are subjugated to the "national good." This leads to irrational use of resources, as the cases of the Soviet Magnitogorsk steel combine and Albert Speer's monumental plans to rebuild Berlin will demonstrate. To be sure, decision making about which projects to fund involves give-and-take among engineers, economic planners, and party officials. Naturally, officials stress the interests of the state. Hence engineers in totalitarian regimes tend to be more accountable to the state; those in pluralist regimes find greater autonomy in setting the research agenda. Their professional societies sell expertise, receiving the exclusive right to practice their professions through licenses granted by the state: for example, as medical doctors. In totalitarian regimes, societies, clubs, and associations for architects, scientists, lawyers, and doctors are subjugated to single-party organizations ruled from above.

Third, technologies in totalitarian regimes are characterized by gigantomania: for example, Speer's plans for wide-gauge (four-meter) railroad track with two-story-high cars or Stalin's seven "wedding-cake" Moscow skyscrapers and the world speed and distance records that were set in aviation (Bailes 1978, 381–406). This scale concerns both physical parameters and the display value of the technology. Gigantomania often results in waste of labor and capital resources, especially in centrally planned economies where the state is the prime mover behind every project. In totalitarian regimes projects seem to take on a life of their own, so important are they for cultural and political ends as opposed to the ends of engineering rationality.

The Technological Style of the Soviet Union

The Soviet Union embraced large-scale technologies with an energy that belied its economic backwardness.[1] Its leaders saw technology as a means to convert a peasant society into a well-oiled machine of workers dedicated to the construction of Communism. They believed large-scale technologies would marshall scarce resources efficiently and provide the appropriate forum for the political and cultural education of a burgeoning working class. Soviet leaders had the utmost confidence in the ability of technology to transform nature and bring freedom to Soviet citizens. Constructivist visions of the Communist future found expression in Lenin's electrification, Stalin's canals and hydropower stations, Khruschev's atomic energy, and Brezhnev's Siberian river diversion project. There were glorious chapters in the history of large-scale technology in the USSR, including the pioneering conquests of the atom and the cosmos.

In all of these projects, Soviet engineers and Party leaders took an extremely utilitarian view of the importance of science and technology to secure dominion over nature. This view was central to the works of thinkers as far back as Francis Bacon in the seventeenth century (Bacon 1942). Marxists further believed that natural laws not only existed in nature but could be discovered in human institutions and applied for the betterment of humanity.

The Soviet Union sought modern technology in the West through technology transfer, including "turnkey" factories (supplied ready to work) and other cooperative arrangements, and through espionage. Its leaders were particularly enamored of borrowing American technology, which they considered the most progressive. Most Soviet engineers believed that once technology had been lifted physically and psychologically out of its capitalist environment, it would cease to serve its capitalist masters and work for the good of the Soviet state and proletariat. It is ironic that Ford, General Electric, and other giant American corporations were a model for Soviet planning.

Both engineers and the Communists favored a strong central government. They shared technological goals. When they seized power the Soviets had no blueprint for industrialization or expertise in organizing it, so the technical specialists provided that. Historian Kendall Bailes argues that "the Communist Party supplied the link, largely missing before the revolution, between the masses of the population and the plans and projects of the techno-structure" (Bailes 1978, 415–17).

The 1920s are often seen as the golden age of Soviet society, uplifted by utopian and constructivist visions for the advent of Communism. For Communist leaders, the view of progress was inextricably tied to technological development, which would be achieved in short order, they believed, through economies of scale, centralized economic planning, mass production, and universal mechanization. This in turn would lead to more rational and equitable distribution of goods and services. Of course, the Soviets encountered great challenges in rebuilding their economy after the ravages of war, revolution, and civil war. It was a long time before modern technology penetrated industry.

Although it entered the economy slowly, technology rapidly became a central aspect of Soviet daily life. There were festivals of machines, symphonies of factory steam whistles. Newly married peasants were conveyed in celebration on a tractor (Folop-Mueller 1965; Stites 1989). (The Nazis, too, embraced these public spectacles as a way to gather thousands of believers together to gape at *Wehrmacht* weapons.) In Soviet literature, technology was displayed with utopian fervor. Posters with technological themes supplanted the Russian Orthodox icon. The machine was central to Soviet commercial art. Technology even had an impact on language, as when proud parents named their boys "Tractor" and girls "Electrification" or "Domna" (forge).

The style of Soviet technologies was characterized by an aesthetics based on two concerns. The first was an exaggerated level of interest in mass production, owing both to egalitarian ideological precepts and scarcities of finished goods. The latter contributed to a premature fixing of parameters for many technologies. The second was the gigantomania that grew out of a fascination and commitment to a technology of display. On the surface, some of these characteristics are reminiscent of the Western Bauhaus movement, Fordism, and Taylorism, with their aesthetics based on standardization, rationalization, and mass production of components (Gropius 1965). But the Soviet characteristics conspired with political forces to create a style noteworthy for bland, functional designs in which safety and comfort played a secondary role, environmental issues were rarely raised, and large-scale systems acquired substantial technological momentum. Moreover, there was a progression of the objects of the transforming visions of large-scale technologies from people to nature itself. First, peasants and enemies of the people were to be transformed into workers and citizens. The metamorphosis of capricious nature into something rationally ordered by technology followed.

The Technicism of Soviet Marxism

Ideology reinforced the Soviet cultural embrace of technology. Soviet economic achievements would be founded on science, technology, and Marxism. Marxism, itself considered an objective science, contributed to almost unbounded faith in science and technology as a panacea for social, economic, and political programs. Soviet leaders from Lenin onward embraced this view. A devout pragmatist, Lenin recognized the central place of science and technology in economic development. In *The Development of Capitalism in Russia* (1899) Lenin argued that large-scale production had supplanted cottage industry, as it would capitalism, throwing manual skill overboard, transforming production on new, rational principles, and systematically applying science to production. Lenin, initially ambivalent to Taylorism—in the sense of studying the worker as a sort of perfectible machine—came to see it as useful for the task of the resurrection of industry.

Even more than humans, for Lenin electricity stood at the center of all efforts to turn the Russian empire into a vast, mechanized Communist society (Lenin 1956; Rassweiler 1988; Staudenmaier 1974). Who could forget Lenin's maxim that "Communism equals Soviet power plus electrification of the entire country"? Skeptics responded by noting that the equation would mean that "Communism minus Soviet power equals electrification." With electrification, technological utopianism supplanted Lenin's rationalism. Lenin was perhaps its most eager advocate, and hence the critical bridge between Communism and technologism. Lenin thought that electrification would change the factory into "clean bright laboratories worthy of human beings" and "eradicate the cultural and economic chasm between town and country," while household electric light would ease the life of "domestic slaves." Electricity would modernize Russian agriculture, provide grain for export, and turn peasants "into full members of the socialist state." (Coopersmith 1992; Wells 1921).

The state electrification plan was not only a utopian transformationist ideal. It reinforced the centralized nature of industrial development so prominent in Bolshevism. Considering the state of their industry and low base-load demands, the Bolsheviks might have chosen smaller power stations. But electrification was a means to transform society socially, politically, and economically. Its advocates discussed issues of control and direction but rarely asked whether resources might be invested in other services such as medical care. A central power network mirrored—and reinforced—centralized political power (Hughes 1983; Coopersmith 1992).

From Taylorism and Fordism to Stakhanovism

Taylorism and Fordism found fertile Soviet soil, only to give way to Soviet transformationist views of human beings in Stakhanovism. Frederick Winslow Taylor, an American engineer, used time-motion studies to determine the most efficient way for workers to accomplish their jobs. Standing in the factory next to the worker, Taylor and his white-smocked followers examined all aspects of the production process: the bending, lifting, and stacking motions, the interaction between worker and tool, and the location of wood, steel, or other material to be worked. They believed they could determine objectively the "one best way" to organize production, expecting that increases in productivity would be shared by worker and manager alike. Managers supported Taylorism because it placed more knowledge, and therefore power, in their hands. Workers saw the time-motion studies as one more effort of managers to extract more labor in less time for less pay, increase control over their labor skills, and raise profits (Aitken 1985).

A. K. Gastev, a writer of proletarian poetry and an engineer, advanced the Soviet form of Taylorism, the scientific organization of labor (*nauchnaia organizatsiia truda,* or NOT). Gastev saw scientific management as a way to bring a new culture of labor to Russia, while at the same time scientifically raising industrial production. In the Gastev system, proletarians would be numbered and classified, with "men and machines merged." "Even words and ideas," Bailes writes, "would come to have precise, technical meanings, devoid of nuance and emotional connotations, and could be plugged in and unplugged as needed." This supremely rational society was certainly Aldous Huxley's *Brave New World* but was captured more brilliantly in Evgeny Zamiatin's dystopian novel *We* (My) (Zamiatin 1924; Bailes 1977; Koritskii 1990; Stites 1989, 145–64; Hughes 1989b, 255–60).

Stakhanovism might better be known as "Red Taylorism." The Stakhanovite movement was a central feature of the effort under Stalin to force increases in labor productivity. Workers were exhorted by political commissars in every factory to set higher and higher norms. The movement was named after Aleksei Stakhanov, a Donets coal miner who achieved miraculous hourly, daily, and weekly tonnage rates and was held up as an example for all Soviet workers to follow. Stakhanovism resembled Taylorism in the attempt to find reserves of labor and capital productivity in existing production. Party leaders used Stakhanovism to undermine the authority of managers, attack technocratic impulses among engineers, find fault with plan targets not set at sufficiently superhuman levels, and prod workers with crude numerical standards (Kaganovich 1936; Siegelbaum 1984, 1988). In Nazi Germany, parallels to Stakhanovism such as the Strength through Joy Movement were also intended to get workers to work hard for state programs without benefiting from higher wages, better working conditions, or access to consumer goods.

As late as January 1936, Sergo Ordzhonikidze, the commissar of heavy industry, placed Gastev in charge of preparing cadres for the Stakhanovite movement, which Gastev enthusiastically endorsed, and nearly a million industrial workers may have been trained in his institute. Gastev, however, rapidly fell from favor, in part because Stakhanovism was an attack on his technocratic views. He was purged in 1938 and died in the camps.

Amerikanizm: Value-Neutral Technology?

The Soviet Union thus had an ambivalent relationship with America and its technology. Party figures, scientists, and engineers who traveled to America reveled in the glow of its technology—its skyscrapers, public transportation, automobiles and highways, industrial laboratories, and in particular its factories, which they saw as the key to the Soviet future. The Soviets imported thousands of Fordson tractors, in spite of problems with spare parts and fuel. The Magnitogorsk steel mill was modeled on the U.S. Steel Company's plant in Gary, Indiana, and the Gorky automobile factory was built by engineers and workers from Ford and modeled on their River Rouge plant in Detroit, at that time considered the most modern of its kind in the world.

Unfortunately for Soviet plans, technology by itself did not transform people. Unskilled workers, improper use of machinery, and mistrust of the bourgeois specialist and Western experts plagued these projects everywhere. Large-scale projects overwhelmed the best efforts of Soviet planners. These problems were clear from the attempt to build Magnitogorsk. Founded in 1929 at the southern end of the Ural Mountains near vast ore deposits, the Magnitogorsk iron and steel complex would be the largest, most advanced in the world, its operations planned to the smallest detail to ensure clock-like precision. Such problems of capitalist production as labor-management disputes, worker safety concerns, and poor living conditions would be left in Gary, Indiana. But owing to both the nature of the technology and the peculiarities of Stalinism, Magnitogorsk never fulfilled planners' or workers' expectations. Instead of serving as a beacon of enlightened Communist Party rule, the plant now symbolizes all that was wrong with Soviet technological style. Its inhuman, gigantic scale led to inefficiency, dangerous operation, and a failure to produce at the planned level (Rassweiler 1988; Dalrymple 1963; Miller 1970; Kataev 1976).

Life in Magnitogorsk in the 1930s is vividly portrayed in American worker John Scott's memoir, *Behind the Urals* (originally published in 1942). At Magnitogorsk Stalinist centralized political control, large-scale industrial organization, and constant exhortation of workers rarely forestalled such problems as long lead times, poor labor discipline, and failure to meet output targets. Party and trade union organizations were powerless to push construction schedules in the face of material shortages and technological backwardness. Many of the workers were "enthusiasts," but they were also inexperienced and unskilled, they wasted supplies and ruined equipment they did not understand how to use. The Party tried to organize on-the-spot technical education, but its schools were as much for ideological indoctrination as vocational training. Hundreds of enthusiasts, inspired to headlong labor that risked industrial accidents, were maimed or killed on the job.

The socialist worker lived no better than his capitalist counterpart. Magnitogorsk laborers suffered through frigid winters, muddy springs, and garbage-laden streets year round. Even the blast furnaces froze in the winters. Open latrines sat right next to crowded barracks and tents; housing, stores, and schools were only beginning to meet demand by the late 1930s. Everywhere in the USSR large-scale technologies such as Magnitogorsk serve as reminders of what happens when the state is the prime mover behind technological development and when centrally dictated economic, political, and technical concerns predominate over human and environmental ones (Scott 1989; Hughes 1989b, 278–84; Kotkin 1991). Of course, there are many examples throughout the world of unsuccessful state planning and execution of technological development. But the scale of human and environmental failure in the USSR seems rivaled only by that in another totalitarian regime, the People's Republic of China (Feshbach and Friendly 1992; Smil 1993).

So distraught was Stalin with this turn of events that although he had declared in a fit of pique, "Technology decides everything!" he had to change his tune to "Cadres decide everything!" For in his eyes they were deliberately destroying, or "wrecking," equipment. Workers, peasants, and bourgeois specialists alike encountered the charge of "wrecking" for failure to achieve targets or for unintentional damage to machinery. The Soviet press portrayed these unfortunate individuals as a kind of Luddite allied to hostile Western interests. The Party held a series of show trials—the Shakhty, Industrial Party (Promparty), and Metro-Vickers affairs—to skewer these *kulaks* (prosperous peasants), *vrediteli* (wreckers), or *spetsy* (specialists), all of which were pejorative terms (Kuromiya 1988).

While *Amerikanist* faith in the utilitarian universality of technology faded under Stalin, and especially with his establishment of economic autarky, faith in technology itself did not abate. The vision of technology as the highest form of culture found its supreme manifestation in Stalinism.

Stalinism, Hero Projects, and Gigantomania

Administration, organization, and technological hubris were joined in the Stalinist utopia. Stalin's views of the place of technology in society were far more economically determinist and far less subtle than Lenin's. This means that the development of the productive forces was the sine qua non of Stalinism. Centralized political control, organization, and economic management were basic to this endeavor, the means for identifying labor and capital reserves and overcoming any human, natural, or technical obstacles. Such obstacles were labeled as evidence of "wrecking." No scale or tempo seemed impossible; everything qualified as superlative, as the "best" or the "biggest" in the world. The most modern technology would secure a safe haven for "socialism in one country."

The Soviets had come to power without a coherent urban policy, but by the end of 1918, after they nationalized land and then abolished private housing in cities, the state became the single client for large-scale construction projects. This gave planners unprecedented control over the urban environment. Soviet architects of the 1920s debated nearly every urban and architectural issue. Would the Marxist utopia consist of nodal points? Would cities be built in a linear fashion adjacent to power and transportation corridors, a style called "automobile socialism" by its detractors? (Lissitzky 1984; Kopp 1970; Ruble 1994).

These questions were resolved in Stalinist fashion in Moscow. On the eve of the Great Break, Moscow's demographics required a change in the haphazard planning that had existed to that time. There was high unemployment, peasants were streaming to cities, and white-collar employees emerged as the largest occupational group, often as representatives of the growing Bolshevik bureaucracy. The city was overcrowded and short on housing (Ruble 1994, 357). Moscow could no longer survive as it was. The Party called for the socialist reconstruction of Moscow as a showcase for socialism's achievements. The center was to be rebuilt for technological display: river embankments were faced in granite; the Volga-Moscow canal was dug (using slave labor) to secure Moscow's centrality in a growing economy; a new Palace of Soviets was designed that would have been three-quarters of a mile tall with a 260-foot statue of Lenin at the top; and the Moscow subway, or metro, was built.

The first technology to serve as the flagship of the Stalinist system was the Moscow subway. This was part of the plan to transform the urban landscape in a socialist fashion. Moscow lagged far behind the cities of Europe and the United States in terms of transportation, hot and cold running water, heating, electric supply, and sewers. Socialist reconstruction, as it was called, would result in well-illuminated streets, parks, and transport systems that returned the worker home rested, not exhausted as in the West, and with the appropriate political education (Kaganovich 1931). In contrast to the dirty, damp, and dark subways in capitalist systems, the socialist metro would invigorate the workers' spirits with its modern ventilation. The aesthetics of the subway stations attracted attention, too. The architecture, sculpture, and paintings of the ornate marble and granite underground palaces reflected dominant themes of Soviet culture: industrialization, collectivization, literacy, and, later, the victory over the fascists: "Each of these palaces burns with one flame—the flame of fast approaching victorious socialism" (Kattsen 1947; Makovskii 1945).

The Moscow metro became the exemplar of future Soviet large-scale technologies. The Party forced the pace of construction against all technological challenges with centralized control, borrowing some techniques from the West but always striving to demonstrate that the Soviet way was better. The method involved serial production at factories of large components, which were transported by rail to site for assembly. Valuable resources were extracted from the rest of the empire, in this case marble, granite, and labor for the "grandiose constructions" (Makovskii 1945). An intimate of Stalin's and one of the few members of the Politburo to outlive him, Lazar Kaganovich, who was in charge of Moscow's socialist reconstruction, proclaimed that the subway "far exceeds the bounds of the usual impressions of technological construction. Our metro is a symbol of the new society which is being built." The metro was also a symbol of the victory of Bolshevik organizers over nature (Kaganovich 1935).

According to a radical plan adopted in 1935, the center of Moscow would be razed "in an effort to 'rationalize' the new socialist capital." There would be new parks and residential areas, but more attention was paid to the "socialist heart" of Moscow. Huge government buildings would be erected to give workers a feeling of the overpowering authority of the regime. Radial highway arteries would converge at the center of the city, with avenues widened. The 1935 plan was not realized because of disorganization, material shortages, and an undercapitalized construction industry, in part because economic ministries organized around sectors of industry had priority for resources over municipalities. Further, building plans transcended the technological capacities of builders. Only a few radial roads were ever built. The Palace of Soviets had to be abandoned as foundations kept sinking into the mud of the Moscow River floodplain. The Nazi invasion and then Stalin's death saved Moscow's Red Square from "tragic disfigurement." But Stalinist planners succeeded in razing one thousand buildings dating from the fourteenth through the nineteenth centuries (Ruble 1994, 362–68).

We see the impact of centralized planning and overemphasis on development of large-scale systems, and capital in general, in the significant human and environmental costs of the Three Gorges Dam project on the Yangtze River in China. The water backing up behind the Three Gorges Dam will inundate tens of thousands of square miles of land, lead to the dislocation of millions of Chinese peasants, and contribute to pollution.

Technological Momentum in Totalitarian Systems

Centralization, bureaucratization, an economically determinist philosophy of technology, and Stalinism combined to give great technological momentum (Hughes 1969; Teich and Lambright 1976–77) to the large-scale systems that were paradigmatic for the Soviet Union. Technological momentum refers to the tendency of large-scale projects to acquire significant social, political, and economic support and of the organizations involved in their construction to become intolerant of obstacles to their diffusion: for example, public opposition. In the absence of market forces, once Soviet construction organizations fulfilled their initial purposes they seemed to take on a life of their own. In a market system, workers and firms might temporarily be displaced but would move to new areas as construction demanded. In the USSR, in order to avoid unemployment and investment in transportation infrastructure, housing, or equipment, Soviet planners sought to provide funding for projects that made use of workers already employed and organization and equipment already in place. Such an approach was needed in the Soviet system to distribute workers because there was no other efficient means. The state and its organizations provided housing, schools, day care, and stores. Indeed, workers preferred working for large organizations with these resources, even though this discouraged their mobility. When projects were finished, institutions and people were transformed into solutions looking for problems.

Technological momentum contributed to premature standardization. The construction of large-scale technologies—subways, hydropower stations, and apartment complexes—required huge capital outlays. Early standardization was a simple way to reduce capital costs. It was no easy matter to establish the specifications for all of the subsystems involved: steel piping, conduit, wiring, prefabricated concrete forms, motors, turbogenerators. Initially, questions concerning construction of large-scale systems required ad hoc decisions by the engineers and managers. Fear of "wrecking" charges made them risk-averse. The engineers naturally gravitated toward accepted practices and norms, rather than innovation, and came to believe one model was enough for the entire empire. Scientific organizations subordinate to a specific branch of industry also focused efforts on increasing short-term production, turning to standardization of component parts. Proletarian aesthetics also contributed to the utilization of fewer redundancies in construction, for example, thicker-gauge pipe or containment vessels for nuclear reactors. The Chernobyl reactor explosion (1986) and the Usinsk oil pipeline disaster (1994–95) serve as reminders of the Soviet technological legacy (Josephson 1986).

Soviet Technology: Proletarian Aesthetics

Proletarian aesthetics grew out of the effort to find economies everywhere in huge engineering projects. In construction, it led to the adoption of simple, prefabricated concrete forms for apartments, offices, and highways. Soviet factories had a universal style, which employed corrugated steel roofs and standard piping, conduit, generators, and machine tools. Engineers in Gosstroi SSSR, the central state construction commission located in Moscow, established codes and specifications for all building materials for the entire empire, irrespective of local geological, meteorological, and other considerations. Industry received appropriations for operations but little for repair and monitoring, making safe operation of these systems a nearly impossible task. One of the most significant manifestations of proletarian aesthetics was shared rights-of-way, where pipelines, highways, railways, and electrical power lines occupied the same thoroughfare. Shared rights-of-way contributed to such disasters as the gigantic Ufa pipeline explosion, which obliterated more than a square kilometer near Tobolsk, Bashkiria, in 1989 and killed more than six hundred people.

Similar kinds of standard techniques were applied universally in the construction industry (Moiseenko et al., 1990). Apartment buildings, subways, even the street names (Lenin, October, Revolutionary, Red Banner) were so similar throughout the USSR that you could fall asleep in one city, wake up in another, and not know the difference. As big concrete slabs replaced bricks, problems of aesthetics began to rival those of quality. The large panels produced not apartment buildings but gray houses of cards stacked on top of each other at right angles. It was not surprising that such buildings as these collapsed instantly during earthquakes in Armenia in 1988. Planners and engineers had hoped that the application of these mass production techniques and materials might enable them efficiently to overcome the poor materials and workmanship endemic to the Soviet experience. They overestimated the ability of mass production construction techniques using prefabricated concrete forms and mass-produced slabs to overcome poor workmanship.

In a word, gigantomania, display value (industrial symbolism in competition for prestige with the West), and state control and centralization of R and D characterized technologies in the Soviet Union. So did fascination with economies of scale and mass production. Did Nazi Germany share any of these characteristics?

Weimar Culture and Technology

National Socialist ideology was ambivalent about modern technology. On the one hand, technology was central to efforts to rearm Germany and secure the new empire's glorious future. The superweapons its army leaders sought during the war required the input of technological experts. Its Four-Year Plan, adopted on the eve of World War II to prepare the economy for the *Blitzkreig*, was an agglomeration of macroeconomic techniques and state-supported projects geared to produce a great industrial power. Many engineers welcomed the strong central government of the National Socialists for its ability to support modern technology more efficiently than the Weimar regime, which had been plagued by an inexperienced liberal parliamentary government and the chaos of the free market. In 1914, imperial Germany had been the leading scientific and industrial nation in the world. Its engineers saw the new chemical weapons and airplanes of World War I as signifying the glories that might be achieved by wedding technical knowledge to a strong state power (Herf 1984, 152–62).

On the other hand, German engineers on the whole were conservative individuals who rejected the rationality of Enlightenment social progress. They believed in the ability to understand physical processes empirically, but not in the extension of empirical methods to human problems. Nazi ideologists, for their part, detested the modern symbols of Weimar technology. They perceived in its spare, utilitarian architecture "Bolshevist" designs, which abandoned the natural antiurbanist aesthetic that should have characterized *völkisch* technology. Nazi Germany needed modern technology to achieve its imperial ends but rejected this technology on anachronistic ideological grounds and found great fault with Soviet technological style. Yet the Nazi technological style—characterized by a preference for centralized project management with attendant social and political control and by gigantomania—was paradigmatic for totalitarian regimes, as I explore in the following sections on Weimar and Nazi architectural achievements, the reception of scientific management (Taylorism), Fordism, and Americanism, and such big technological projects as the V-2 rocket.

The hyperinflation, unemployment, and leftist and rightist uprisings of the early Weimar years gave way to economic prosperity and economic stability in the late 1920s. Efforts to employ Fordism and Americanism for higher productivity and industrial rationality reached their zenith. But as the Great Depression of the early 1930s set in, the political parties on the extremes grew, with the Nazis the ultimate beneficiaries. Throughout the period, the conservatives blamed the lost war, the unfavorable terms of the Versailles treaty, and economic instability on Jews and liberals. They viewed cosmopolitan mass culture as decadent and attacked political liberalism as "Bolshevist" (Herf 1984, 18–22). The engineers among them saw the left as promoting technological stagnation through their fear of rearmament and criticized liberal political and social programs for promoting "cultural Luddism"; this meant they believed the liberals would destroy *völkisch* culture as the anti-industrial Luddites of early nineteenth-century England had smashed machines to bits. National Socialism gave them hope of fulfilling self-interest and state service, unleashing technology from the bonds imposed by the Weimar Republic. Engineers eventually talked themselves into believing that the Nazis had abandoned the anti-industrial themes of *völkisch* thought (Herf 1984, 152–62).

What was it about Weimar that raised the specter of Bolshevist technology? To many German engineers such vibrant cultural phenomena as the Bauhaus, an artistic and technological movement born in the Weimar Republic, symbolized everything that was non-German about modern technology. Like other modernist architects such as Le Corbusier and Frank Lloyd Wright, Bauhaus architects transformed the house or other building into a tool. They sought to integrate craft, art, and industry in one modern aesthetic. They appealed to such ideals as democracy, optimality, and efficiency in their designs. For Le Corbusier, the park and the skyscraper were united to exalt rational power. Wright, in his utopian vision Broadacre City, saw the telephone and automobile as contributing to the disappearance of the city because these technologies were inherently democratizing in their decentralizing force and their ability to diffuse population, wealth, and power (Fishman 1982; Whitford 1984; Banham 1986). The democratic ends of such architecture disturbed the conservative Germans.

Bauhaus artists such as Walter Gropius and Ludwig Mies van der Rohe seized upon the aesthetic of the machine, its embodiment of speed, efficiency, and clean lines, in every thing they produced: office buildings and apartments, chairs and other furnishings, even utensils and vases. They took inspiration from factory design. They sought to bring together artists, craftsmen, sculptors, and architects in the common endeavor of tying crafts to industry and mass producing craftwork. Architects secured funding from the Weimar and municipal governments for many of their projects for mass housing and concentrated on the technical aspects of construction technology and planning. Some of them claimed "that construction methods determined style itself," with standardization of building parts—prefabricated concrete forms and modular construction—essential for a uniform aesthetic. They believed that mass housing was not different from mass transportation or any other problem of urban planning (Lane 1985, 128–30). Bauhaus supporters believed that technology could be employed to achieve a diversity of modernist social ends—inexpensive housing, rapid mass transit, etc., and their implicitly democratic ends—through standardized means. No matter what Bauhaus architects believed, socialist, modernist, or totalitarian ends can be achieved through the application of standardized construction techniques.

Most German engineers, however, believed that this style of technology could not be reconciled with German culture in a nationalist ideology. In spite of the fact that the Bauhaus style was recognized internationally as an achievement of German culture, for conservative elements it was "un-German," based on a uniform industrial aesthetic, a proletarian social policy, and a "Bolshevist" political program of helping the masses. They attacked its prefabricated housing and standardized building methods. Conservative architects appealed to national building traditions. They considered that flat roofs, for example, provided inadequate draining of rain and melting snow and were inappropriate to the German climate; according to one, the flat roof was an "oriental form" and equated with flat heads. Other architects declared that standardized building techniques produced "'nomadic architecture,' leading to 'uprootedness, spiritual impoverishment and proletarianization.'" Still others adopted an antiurban theme, attacking skyscrapers and calling for a return to the German soil. Eventually, this kind of criticism of the Bauhaus was incorporated into racial arguments, where the origin of the Bauhaus style was attributed to cultural decadence that had its roots in biological causes. These kinds of explanations, of course, won support in Nazi circles (Lane 1985, 132–40).

Nazi Technology

Were National Socialism a consistent ideology, we would expect efforts to create an agrarian society in which the *Volk* could best prosper. But the Nazi rise to power did not give way to rural nostalgia or to an antimodernist technological ethos that supported the peasant's organic tie to the soil. Rather, Nazism combined anti-Semitism and the embrace of modern technology in a myth, according to which technological advance grew out of a racial battle between Aryan and Jew, blood and gold. The engineer would assist the regime in destroying an unhealthy urban atmosphere, liberating the nation from the "fetters of Jewish materialism. The Nordic race was ideally suited to use technology; the Jew misused it." Like Soviet, Nazi technology embodied service to the nation, not pursuit of profit. Service to the nation meant joining with the state to achieve economic independence. State trade, tariff, tax, price, and wage policies would help to underwrite technological development to achieve autarky and enable the nation to engage in war when cut off from the import of raw materials (Herf 1984, 189–93).

For Hitler himself there was no *völkisch* rejection of technology. If in life and politics the strongest won, so among nations the technologically weak would be defeated. Hitler advocated rearmament and, like Stalin, national autarky. He used new media such as radio and film to praise *völkisch* technology for propaganda ends and sponsored modern highways—the autobahn—and other modern artifacts for economic and military ends (Herf 1984, 194–96). And when the war effort bogged down, Hitler hoped for a technological savior in the form of a new superweapon like the V-2 rocket.

Hitler's writings and speeches criticized Weimar culture for its weaknesses, its decadence, its materialism, its "lack of an heroic ideal," its "Bolshevist" art. He singled out its architecture as the epitome of these problems. Hitler supported gigantomania in Nazi architecture. The Nazi leaders built massive monuments to their rule whose neoclassical style and scale were neither *völkisch* nor humanistic. Hitler believed that a "great" architecture was needed in the Third Reich, since architecture was a vital index of national power and strength. There had to be monuments in cities, not symbols of cultural decay.

When the Nazis began vigorously to oppose the Bauhaus in the 1930s, the foremost party philosopher, Alfred Rosenberg, led the charge. Rosenberg had joined the Nazi Party in 1920 and became editor of *Völkischer Beobachter* in 1921, through which he attacked the Bauhaus. His career had ups and downs, but in 1933 he was put in charge of ideological training of Nazi Party members. From this position, Rosenberg hoped to see the Nazi Party university create a place for natural science, especially to study the biological laws of races to reveal the poisonous influence of the Jews. In 1941, Rosenberg became the Reich's minister for the eastern occupied territories, a position from which he could see Nazi racist policies of *Lebensraum* and the "final solution" put into action. He was hanged for war crimes in 1946 (Rosenberg 1970; Pois 1970).

In 1929, Rosenberg founded the Kampfbund to spread a Nazi gospel of virulent Christian anti-Semitism and racial doctrines. Many of these doctrines were based on the writings of Count Gobineau, who had argued that the rise and fall of civilizations is connected to their racial composition. Those of Aryan stock flourish, while those diluted through miscegenation decline. Rosenberg embraced conspiracy theories and feared the "international Jew," freemasonry, and Jewish control of banking and the media. The Kampfbund was central in spreading Rosenberg's message of the *völkisch* aesthetic. Initially, the Kampfbund had the strong political backing of the Nazi Party. In the same way that Bolshevik organizations had subjugated engineers' professional associations, so the Kampfbund inexorably absorbed smaller rivals. The Kampfbund set forth the party line on cultural values. It attacked the chaos, Russian "Bolshevism," and American "mechanism" allegedly rampant in modern art, the Jewish roots of these problems, and the "Nigger-Culture" that thrived in the Weimar clubs whose excitement and decadence was so well captured in the Broadway musical, and later film, *Cabaret* (Lane 1985: 149–51). Rosenberg despised modern art. He saw in Picasso "Mongrelism," whose "bastardized progeny, nurtured by spiritual syphilis and artistic infantalism was able to represent expressions of the soul"; and he hated the work of such artists as Marc Chagall and Wassily Kandinsky, who was connected to the Bauhaus (Rosenberg 1970, 128–51).

Rosenberg argued that Bauhaus architecture was a symbol of weakened culture, of a mass society whose members had lost their historic identity through urbanization and their economic security through proletarianization and unemployment. The Bauhaus was a "cathedral of Marxism" that resembled a synagogue or a "Bolshevist" building for the nomads of the metropolis (Lane 1985, 162–63). Rosenberg constantly referred to the interconnection of race, art, learning, and moral values in his attacks on the cultural decadence of Weimar, attacks that, to the German people who were suffering through the depression, were appealing (Lane 1985, 147–49). On April 11, 1933, the Berlin police shut down the Bauhaus school by order of the new Nazi government.

Next the Nazis orchestrated the *Gleichschaltung* of municipal building administrations and building societies by purging them of adherents to the modernist Bauhaus movement. The societies were then joined in a central organization under government control, just as all professional societies were subjugated to Communist Party organs under Stalin. In spite of their criticism of "Bolshevist" urban development, the Nazis supported programs for large-scale, low-cost public housing in their appeal for working-class support. This support of public housing resulted in part from Rosenberg's loss of influence to Joseph Göbbels. The Kampfbund was placed under Göbbels's authority, and he established the Reichskulturkammer as a branch of his Propaganda Ministry (Lane 1985, 169–71). The Kulturkammer had sections for film, literature, theater, music, the media, and visual arts, with national and regional offices. Göbbels himself hesitated to purge the new style entirely from the Third Reich. Barbara Lane writes: "If the establishment of the Reichskulturkammer cut short the purges of 1933 and prevented the original leaders of the Kampfbund from gaining control of architectural style, Göbbels's organization never explicitly repudiated the Kampfbund's attacks on the new architecture; and these attacks had a profound effect upon the careers of the radical architects" (Lane 1985, 176–84). Many modernist architects were deprived of their livelihood and had to emigrate, like physicists, biologists, and doctors. While depriving Bauhaus architects of influence and dictating issues of style, this did not prevent them from getting new commissions, and many Bauhaus assistants and students received positions in the Nazi government (Lane 1985, 171–73).

Gigantomania in Nazi Germany

Nazi architectural style, like Soviet, was gigantomanic. Hitler desired immense monuments to his rule and the glory of the Third Reich for millennia to come, buildings of a scale never before seen. The party and its strong central state were the driving force of Nazi architecture. Nazi buildings were intended to express the will of the Nordic people, awaken national consciousness, and contribute to the political and moral unification of the *Volk* (Lane 1985, 185–89). In October 1935, when the major structural frame of the new Luftwaffe building was finished, Hermann Göring addressed gathered workers and functionaries to praise the structure as "a symbol of the new Reich," a building that "shakes our deepest emotions," shows German "will and strength," and would "stand forever like the union of the Volk" (Göring 193, 148–50).

Nazi architecture was not one, historicist style, nor an out and out rejection of the new style, but a variety of styles that reflected the diversity of views of the leadership: public works such as highways and bridges, government buildings, and some apartment buildings; neo-Romanesque; rustic housing projects intended to tie the urban workers to the soil; modern neoclassicism based on the Doric aesthetics of Albert Speer; and even modern (Speer 1970, 62–63; Lane 1985, 185–89).

Albert Speer was the chief architect behind many of the gigantic projects. Hitler desired Speer to make a huge field for military exercises and party rallies, with a large stadium and a hall for Hitler's addresses. While never completed, the planned Nuremburg tract embraced an area of 16.5 square kilometers (roughly 6.5 square miles). All of the structures would have been two to three times the size of the grandiose Greek and Egyptian constructions of antiquity. For example, Speer designed the Nuremburg stadium based on the ancient stadium of Athens, but far larger: six hundred yards by five hundred yards. Speer selected pink granite for the exterior, white for the stands. To the north of the stadium a processional avenue crossed a huge expanse of water in which the buildings would be reflected. When Hitler first saw the designs he was so excited, his adjutant reported, that he "didn't close an eye last night" (Speer 1970, 64–67).

In 1939, in a speech to construction workers, Hitler explained his grandiose style: "Why always the biggest? I do this to restore to each individual German his self-respect. In a hundred areas I want to say to the individual: we are not inferior; on the contrary we are the complete equals of every other nation." Hitler's "love of vast proportions," Speer commented, was connected not only with totalitarianism but with a show of wealth and strength and a desire for "stone witnesses to history" (Speer 1970, 69). Yet these structures could scarcely have instilled in the individual any personal feeling other than insignificance. For anyone but Hitler himself, any sense of glory could come only as an anonymous contributer to the all-powerful state.

Hitler wanted a new chancellery to celebrate his rise in rank to "one of the greatest men in history," with great halls and salons to make an impression on visiting dignitaries. He insisted that it be built within a year. Speer was required to raze an entire neighborhood of Berlin. Forty-five hundred workers labored in two shifts, with several thousand more scattered throughout the Reich producing building materials and furnishings. To meet Hitler's designs, Speer created a great gate, outside staircase, reception rooms, mosaic-clad halls, rooms with domed ceilings, and a gallery twice as long as the Hall of Mirrors at Versailles. The chancellery included an underground air-raid shelter. When it was finished, Hitler "especially liked the long tramp that state guests and diplomats would now have to take before they reached the reception hall" (Speer 1970, 102–3, 113–14).

The future headquarters of the Reich would have been the largest structure of all, with a volume fifty times greater than the proposed Reichstag building. It could have held 185,000 persons standing and was "essentially a place of worship." Its dome opened to admit light. At 152 feet in diameter, it was bigger than the entire dome of the Pantheon (142 feet). A three-tiered gallery was 462 feet in diameter and 100 feet tall. In order to ensure that the structure lasted into the next millenium, engineers calculated, its steel skeleton, from which solid rock walls were suspended, would have to be placed on a foundation of 3.9 million cubic yards of concrete, dozens of feet thick; the engineers did tests to determine how far the monstrous cube would sink into the sandy building site. Hitler was partly motivated by Stalin's projects. "Now this will be the end of their building for good and all," Hitler boasted (Speer 1970, 151–55).

Hitler desired to rebuild Berlin as the capital of "Germania," a new empire that would span the entire Eurasian continent and far outdistance Rome, London, and Paris in grandiosity and history. Hitler had studied the Ringstrasse in Vienna with its prominent public buildings (Schorske 1979). Speer had to order the heart of the city razed to accommodate the two new axes through the center lined with tall office buildings. Four airports were situated at the terminal points of the axes. A ring Autobahn encircled the new Berlin, incorporating enough space to double the city's population. A four-story copper and glass railway station with steel ribbing and great blocks of stone, elevators, and escalators would surpass Grand Central Station in size. The plans themselves experienced gigantomania, eventually including seventeen radial thoroughfares, each two hundred feet wide, and five rings; the land beyond the last ring would be for recreation, a woodland of artificially planted deciduous trees instead of indigenous pine. The projects required immense effort; SS head Heinrich Himmler offered to supply prisoners to increase production of brick and granite, which were in short supply. Himmler's SS concentration camp operations showed tremendous ignorance of construction techniques and often produced blocks of granite with cracks. They could supply only a small amount of the stone needed; highway construction used the wasted material as cobblestones (Speer 1970, 73–79, 134–35, 144). Only the demands of war prevented the Nazis from carrying out these radical transformation plans.

Hitler's favorite toy, it seemed, was the model city, a 1:50 scale model that was set up in the former exhibition rooms of the Berlin Academy of Arts. When Speer's father saw the mock-ups he commented, "You've all gone completely crazy." Only later, when in prison, did Speer realize the inhumanity of his designs, the "lifeless and regimented" nature of the avenues, the "complete lack of proportion" of the plans (Speer 1970, 132–39). But we should not think of Speer's designs as unique in the Western world at the time. There was a resurgence of interest in massive neoclassical forms in other Western countries: for example, Rockefeller Center in New York City, the forty-four-story gothic Cathedral of Learning in Pittsburgh, and Stalinist architecture generally.

The heroic Nazi projects pushed to the limits of technology. Very few large projects were carried out, in part because of their astronomical costs and the costs of war. Smaller, more feasible projects became showpieces of Nazi propaganda, with Hitler a prominent figure at groundbreakings. There was constant coverage of the projects, some of which took years, and this propaganda all but obscured the failings of the building program: for example, projects for the masses such as public housing never met demand. Nazi public housing retained the Weimar (and universal) practice of constructing row houses and apartment buildings on the periphery of urban population centers. Only a few projects conformed to the Nazi ideal of tiny houses with sloping roofs, sited on enormous plots of land. The surfacing of these attempted to evoke the countryside: thatched roofs, half-timbering, or vertical wood siding (Lane 1985, 190, 205–13).

There was a contradiction between the designs of the Reich's commissar for public housing to fill the world with lovely peasant houses in the postwar period and the plans of Speer as general building inspector for Berlin to undertake its rebuilding as a "world capital city" of "insane monumentality" whose buildings would be an "imperishable confirmation" of the power of the Third Reich, yet had a "nonstyle of pseudo-antique form, ponderous excess and solemn emptiness." But Hitler recognized this. Referring to his government's new palace he said, "Amid a holy grove of ancient oaks, people will gaze at this first giant among the buildings of the Third Reich in awesome wonder" (Bracher 1970, 347). But this contradiction displays a central contradiction of totalitarianism itself: the superefficient omnipotent state constructed of people kept in isolation and ignorance.

Economic Planning, Scientific Management, and Autarky

All governments use economic indicators to develop short- and long-range plans. In market economies, governments prefer to allow market mechanisms—millions of individuals acting on prudent economic self-interest—to determine wages, prices, and the allocation of goods and services. The interests of business in higher profits and lower costs are supposed to ensure the pursuit of innovation and efficiency naturally. But governments turn to such measures as taxes, tariffs, subsidies, and regulations to assist the market in achieving such goals as full employment, higher productivity, and health and safety measures. For as the case of the United States shows, the belief that government ought to remain distant from all macroeconomic policy is held only by a minority. The government performs useful functions from which all citizens benefit, such as gathering and publishing statistics that help businesses make their decisions, and its regulatory and other agencies protect the environment, ensure workers' employment security and occupational safety, and prevent monopolies, if not oligopolies, from forming.

In the centrally planned economies that are characteristic among authoritarian regimes, the government's role is far more central in shaping economic activity. In the socialist model of the USSR and Eastern Europe, the state owned the means of production, indeed most property, was essentially the sole employer, and directed the production process through massive bureaucracies that set prices and wages, allocated resources, and established priorities. Trade unions with the power to oppose monopolies through collective bargaining were subjugated to state organs. Since new technology requires enormous investments with great risk and uncertain returns, only the largest corporations (in Nazi Germany) or the state itself (in the USSR) could underwrite these endeavors. Investment and profit, foreign trade, and labor all become the purview of state (party) planning organs.

In authoritarian regimes, economic planners and party officials believe they can employ the state's power to overcome the weaknesses present in market economies. In Stalin's Russia this meant eliminating the profit motive, ending the exploitation of the worker, and ensuring that the basic living needs of each individual were met. In Hitler's Germany, the goal was harnessing the economy to the military and imperial designs of the state and defeating the hated international conspiracy of Jewish capital. Both countries experimented with scientific management techniques such as Taylorism and were attracted to aspects of American industrial organization, especially its use of the assembly line and mass production, which was often referred to as Fordism.

Taylor's doctrines were widely known in Weimar Germany; his works had been translated into German even before the war. Weimar engineers and managers were interested in the promise of scientific management for industrial planning and saw Taylorism as a way toward economic recovery. They admired America's highly rationalized production, standard of living, and functioning democracy. While German workers, like their American counterparts, feared Taylorism for deskilling them and putting more power in managers' hands, German engineers assumed that America's prosperity and the ability of its workers to own their own houses and cars resulted from successful application of scientific management (Hughes 1989b, 284–86).

Henry Ford, anti-Semite, racist, and rabid anti-Communist, developed a large following in Germany, and his system was the rage the world over, even in the USSR. German intellectuals saw appeal in Fordism because the higher wages made possible by Ford's approach would stimulate higher consumption, thence higher production, and then still higher wages. They reveled in the order of his River Rouge production facility. Like Soviet intellectuals, many of them wrote books about their trips to America's industrial heartland, where Ford's River Rouge plant was the most well-known stop (Hughes 1989b, 288–92). Ford and the Nazis also engaged in mutual admiration for their hatred of Jews and Bolsheviks.

Some German engineers tried to separate love of technology from *Amerikanismus*, the American system of mass production based on interchangeable parts. These individuals gained the upper hand in Nazi ideological pronouncements. They believed that "Americanism" was an obsession with economics and that mass production and consumption were linked to oppression. They despaired that America's technology, with its Fordism and Taylorism, was "soulless" and decadent and could never find an appropriate place in German culture. The *Volk* stood for blood, race, and cultural tradition, as opposed to such dangers as *Amerikanismus*, liberalism, commerce, materialism, parliaments and political parties, and an excess of rationalism. Hence *völkisch* ideology rejected liberal ideas of individual rights, socialist assertions that class conflict prevented the achievement of genuine community, and therefore also unions. It was aggressive in matters of foreign policy. And, of course, it was racist (Herf 1984, 35–36, 163, 185).

In spite of their confused yearnings for *völkisch* technology, German engineers energetically contributed to the development of large-scale systems that closely resembled their Western counterparts, including automobiles and highways, economic planning techniques, and research and development in support of armaments. Fritz Todt, who served a number of positions in the Nazi government, is an example of the paradoxical attitude of National Socialism toward the engineer. Todt (1891–1942) was minister of munitions and of *Reichsautobahnen*. He saw the National Socialist government friendly to his efforts to build the Autobahn, the national highway system that was his long-held dream.

As a government official, Todt sought to involve engineering traditions in state programs. He encouraged technical specialists to work on such practical issues as developing natural resources, finding new energy sources, and decreasing German dependence on foreign raw materials. As head of the Union of German Engineers from 1939 until his death, Todt strived to involve engineers in the creation of a rational armaments economy. In this way, he successfully bridged an abyss between Nazi antitechnological irrationalism and the need for engineering expertise. Göbbels considered him modest and unassuming and referred to the "genial spark plug power of his personality." Todt's death in a plane crash in 1942 left Hitler visibly shaken (Goebbels 1948, 41–43; Heinemann-Grüder 1994, 35–36).

Todt's highway system was a symbol of the bridge between Nazism and technology. Todt believed that in its conception the Autobahn differed sharply from the chaos of Weimar public works projects. "It flowed from a unified *Geist* and represented an artistic effort to give proper form to the German landscape." The highways were "more than an engineering feat, but an expression of the German essence." They were evidence that the Nazis had rescued technology from materialism. Aesthetic criteria had displaced the profit motive. This was a "soulful cultural work." Todt said: "The following are the features that make a road as a totality into an artwork that brings the environment joy through its intrinsic beauty and harmony with the environment: the direction of lines is bound to the land. . . . Construction remains true to natural forms. . . . Workmanship is based on the craftman's principles of building and implantation in the earth." The Autobahn, like the peoples' car—the Volkswagen—meant to drive on it, was "commensurate with the needs of the Volk, not a brutal and soulless image of iron and cement" (Herf: 1984, 204–5; Todt 1932).

Begun in the 1930s under his direction, by 1942 the Autobahn was over a thousand miles long and linked Germany's major cities. It was inadequate to handle rapidly growing traffic and was heavily damaged during the war. Still, even more than rearmament, the automobile and highway were the basis of Germany's economic recovery after the Depression. The Depression led to the consolidation of the industry, allowing Daimler, Mercedes, and BMW to become stronger. Between 1933 and 1938, annual German automobile production trebled to 340,000. While only one-fifteenth of the U.S. output, this growth reflected a general economic recovery and stimulated iron, steel, and construction industries. The Nazi policies of road construction, taxes, and incentives underwrote automobilization and overcame German resistance to the auto that stemmed from a fine railroad system, aristocratic preference for the horse, and inadequate infrastructure. The Nazi government also benefited from the ongoing development of synthetic gasoline and rubber (Overy 1975; Hughes 1969).

The automobile generated interest in Taylorism and Fordism, even from Hitler. The automobile industry in Germany had been a pioneer in flow production methods and modern factory organization, increasingly demanding from suppliers the same levels of efficiency and organization that Taylorism brought to Opel and Ford (Overy 1975). Now Nazi fascination with certain aspects of Fordism found response in Hitler's burning desire to see a car built for the German masses, a *Klein-* (small) or *Volksauto*. Within ten days of his appointment as chancellor, Hitler declared the Nazi intention of building such a vehicle. The Nazis invited the leading auto industrialists—Opel, Daimler-Benz, and Porsche—to design a lightweight, low-horsepower, high-efficiency *Kleinauto*. The chosen design would be produced with government subsidy under the authority of the German Auto Industry Association (RDA). The automakers actually opposed the subsidized *Volksautos* because they would compete with their regular models. Eventually the Gesellschaft Zur Vorbereitung des Volkswagens (Society for the Development of the Peoples' Automobile) was established. It employed American and other foreign experts. The Nazis initiated an installment/layaway plan for the Volkswagen. Around 350,000 Germans invested 280 million marks, even though price and delivery date were not guaranteed. Autostadt, like Magnitogorsk built to serve government industrialization programs, was dedicated in 1938. But with the building of the Westwall fortifications opposite the French Maginot line, construction virtually stopped. Autostadt was converted essentially to military production. The Volkswagen "beetle" would have to wait until the 1950s.

Such projects as the *Kleinauto* and Autobahn signaled a new relationship between Germany's industrialists and the state. Nazism's economic policy, to the extent one existed, involved cooperation with large capital to achieve its ends. Within three weeks of being named chancellor on January 30, 1933, Hitler invited a number of Germany's top industrialists—the heads of Krupps, I. G. Farben, etc.—to Göring's offices to press them for contributions and promised to protect the interests of private enterprise. The industrialists, whose government-subsidized projects had languished, welcomed Nazi attention.

The next stage in rearmament and industrial cooptation was the promulgation in 1936 of the Four-Year Plan, intended to support the future war effort and exploitation of territories to be conquered. Hermann Göring headed the office of the plan, with deputies for specific branches of trade and industry, the most important of which were motor vehicles, iron and steel, chemical, and construction industries. The office sought rationalization in German industry through planning and directives. It acquired key economic fiefdoms for party functionaries, for example, a new Hermann Göring Works. The Göring iron and steel combine reflected the Nazis' efforts to transfer ill-gotten gains from extortion and expropriation of property into legitimate businesses. Between 1937 and 1939, it acquired machine tool, armaments, automobile, railroad, and mining industries.

The Nazi economy quickly became highly monopolistic and cartelized. This meant that all productive capacity was restricted to the end of economic preparedness; new enterprises could not be established even with private capital without the full agreement of the office of the Four-Year Plan. "Inefficient and unreliable" businessmen were eliminated—the handicraftsman and the retailer. Many consumer-goods industries for textiles, leather, soap and chocolate were closed as well.

While the Four-Year Plan promoted the goals of rearmament, self-sufficiency, and engineers in service of the *Volk*, it failed for a number of reasons. First, monopolization, cartelization, and Aryanization deprived the economy of much of its dynamism. Second, after war began Germany ignored technical education for political-ideological training. Adolph Hitler Schools had been established for elementary grades, and *Ordensburgen* (order castles) for higher education, that were intended to turn out a technically and ideologically trained elite (as in the USSR). They were not up to the tasks at hand, focusing mostly on ideology (Speer 1970, 122–23; Herf 1984, 198–203). During the war, many universities nearly closed as all able-bodied men were drafted into the *Wehrmacht*.

Third, while trying to cater to the working class, the Nazis destroyed trade unions and used unemployment to tighten working and wage conditions. The Nazis betrayed the words *socialist* and *labor* in their party's name. Labor lost its rights to organize, strike, and bargain collectively. A German Labor Front was organized to represent workers' interests, but it truly was a "front" organization. The German Labor Front goaded workers into higher productivity, offering as a substitute for real workers' rights or increased wages appeals to national pride and the work ethos. Its appeals included the "Soldiers of Labor," "Beauty of Labor (beautification of worksites)," and "Strength through Joy (*Kraft durch Freude*)" propaganda ploys. Obedient German workers paid their Volkswagen installment payments at Strength through Joy and Labor Front offices that carried the VW savings stamps. The ploys resembled Soviet Stakhanovism in the effort to get more our of workers for no more pay. At least, unemployment had abated (Bracher 1970, 331–32).

Similar to Nazi architecture, the Nazi organization of the economy was an amalgam of rational and irrational means and ends. The ends of preparing the economy for war by putting all resources in the hands of the state for efficient use may have been rational by some standards. But the Four-Year Plan exacerbated the raw material situation, which made imperialist aggression for colonies attractive, in turn resulting in the need for more arms. A debate ensued about whether to create in Germany proper the productive center of Europe or to preserve the antiurban ideal Germany while promoting industrial experimentation in the conquered territories. The Nazis relied on experts and economists as instruments and objects, not originators of policy. Short-term efficiency and the primacy of politics, not capitalist or socialist doctrine, determined the course. Though it was intended to be a permanent war economy, Nazi arms production reached full capacity only in autumn 1944, even though augmented by six million foreign workers and slaves (Neumann 1942, 177–79, 249–50, 268–69, 298–300; Bracher 1970, 335, 405).

Göring saw the Four-Year Plan as central to overthrowing the Versailles treaty of 1919, ending the plunder and exploitation of Germany by the Jew, and establishing autarky. He called on Germany's scientists and engineers to use their skills to produce gasoline and mineral oils from coal, new alloys, mighty factories, and buildings for rearmament. He appealed to inventors and scientists for their collaboration. "Think hard, make experiments, work in your laboratories, give us new ideas, new inventions, and new possibilities, and you will have done great things for Germany" (Göring 194, 191–208). But when scientists were involved in these efforts, the results were mixed, owing to the irrationality of Nazi gigantomania.

"Big Technology" in Nazi Germany

The V-2 rocket, the first large guided rocket, was the greatest technological achievement of the Third Reich. Yet it was a poor weapon, unable to carry a large explosive payload, and diverted manpower and other resources from more sensible armaments projects. The V-2 demonstrated the importance of the leader principle in scientific success. Hitler's support set the project off; when he lost faith in the V-2, the program lost priority for material and manpower. Still, the V-2 program foreshadowed the Manhattan Project and "big science" of the postwar years as a paradigm of state mobilization to force the invention of new military technologies. The V-2 program grew out of a military bureaucracy, which rarely considered human or economic costs. This, together with the needs of secrecy and the inadequate technical basis of industry, necessitated the creation of a large government-funded central laboratory (Neufeld 1994, 51–53).

During the Weimar years, a popular fad for rocketry and space flights produced both stunts and serious liquid fuel experiments that generated national pride for the outcast nation (Neufeld 1990). With the Depression the fad ended, but experimentalists continued their trial-and-error efforts. A group that included Wernher von Braun, who would direct technical aspects of the U.S. manned space program, was established in Berlin in 1930. (Von Braun energetically promoted the U.S. program in the 1950s by touting the possibilities of the moon's colonization by millions of people; his colonization claims resembled Nazi *Lebensraum* ideology.) The Berlin group sought corporate financial backing by stressing commercial applications such as intercontinental transport. The turning point in the V-2 program was the interest of the Army Ordnance Office in the use of rockets to deliver chemical weapons. The office built a large secret facility to maintain the assumed lead in rocket development that Germany had over other nations. It used the Gestapo to impose secrecy and to drive other rocket efforts out of business. Secrecy required outside subcontracting to be abandoned and necessitated the fabrication of one-of-a-kind hardware. The program gained momentum when the Luftwaffe (air force) joined in, securing the political support and resources of Hermann Göring. The Luftwaffe underwrote a new weapons facility, Peenemünde (Neufeld 1994, 56–58).

At Peenemünde, von Braun and other scientists sought to build an in-house production line of rocket components for the finished weapons. Peenemünde had the advantage over universities in research and development in terms of concentrated scientific interest, commitment to Nazi or national ideology, stable funding, and draft exemptions for key personnel. Von Braun cultivated contacts with university scientists and engineers for manpower. He successfully created an open academic atmosphere in an environment of secrecy. In this environment "big science" was fostered. When the Nazis invaded Poland, appropriations for research rapidly expanded. Leading military figures associated with the V-2 had laid the groundwork for these increases by promising the deployment of missile weapons in short order and by fostering contacts with such high officials as Albert Speer. But at times Hitler withdrew his support, including cutting deliveries of steel in favor of other priority projects. Even relations with the Luftwaffe deteriorated as the war progressed and Germany's prospects worsened (Neufeld 1994, 59–62).

Speer secured Himmler's intervention in 1943 to push production, utilizing concentration camp labor at newly built underground facilities. In a last-ditch effort to hold off defeat, the Peenemünde facility was pushed to produce as many missiles as possible to use on England. No thought was given to the human cost, in the slaves who toiled without expression to the death in damp, disease-ridden conditions, living, working, and sleeping in the Peenemünde caves (Goebbels 1948, 286).

Hitler's arbitary and autocractic behavior had a negative effect on Nazi "big science." Speer reports that as the military situation deteriorated Hitler made a series of technological blunders: for example, ordering a fighter jet capable of shooting down American bombers to be built instead as a fast but tiny bomber, incapable of holding many bombs. Hitler then insisted that the V-2 be mass produced at a level of nine hundred per month for use as an offensive weapon. Five thousand rockets—five months' production—would deliver perhaps an effective 3,750 tons of explosives; a single combined U.S. and British bomber attack delivered 8,000 tons. But Hitler was determined that some future new weapons would decide the war. So fascinated were the Nazis with a technical fix to their military quandary that they allowed the untried, young von Braun great leeway to pursue the expensive V-2 with only long-term prospects. Yet Speer was also attracted to the romantic possibilities of a superweapon. The Nazis, he later admitted, suffered from an "excess of projects in development," not one of which could ever meet full wartime production, and many of which were rushed "from factory directly into battle" without "customary full testing time" (Speer 1970, 363–70, 409–10; Neufeld 1994, 65–66; Goebbels 1948, 219). The result, of course, was loss of young life. What is characteristically Nazi about the V-2 technology is not this loss of life but the massive build-up of state support behind the project (Neufeld 1994, 70–71).

Technology in Authoritarian Regimes

The Soviet Union had great respect for American technology, even as it despised American capitalism. It was utopian in its embrace of technology, assuming that any technology would function smoothly given socialist economic relations. Nazi Germany rejected Western technology in ideological pronouncements, even as it relied on technology for its military rebirth. It had an irrational, dichotomous attitude toward the role technology might play in the national future. Stalin appears to have been more in tune with the limits and prospects of technology than Hitler.

While these are important differences, several common features of authoritarian technology stand out. In authoritarian regimes, technologies are intended to organize workers into malleable individuals devoted to national goals that divert attention from their economic self-interest: for example, from higher wages. The technologies themselves are large-scale, inefficient, and extremely costly in terms of human lives and the natural environment. Technology has great display value, as the efforts to rebuild Moscow and Berlin demonstrate. The scale of such reconstruction dwarfs people. In its radical design and massive thoroughfares, centralized political power is the message. The huge projects garner more than their share of resources, impoverishing other sectors of the economy. Only technological limitations and war prevented socialist Moscow and Nazi Berlin from being realized.

The careers of engineers in authoritarian regimes, too, suggest parallels. In Nazi Germany the engineers, like the physicists of Planck's ilk, opted for political accommodation rather than resistance, assuming that they could control Hitler or, at least, that Hitler would use "legal" means to achieve his ends. Their accommodation involved assistance in economic planning, industrial management, and armaments, all of which they viewed as service to the nation. In the USSR, unrelenting political pressure forced engineers into service to the state. But Soviet technologists also wanted to serve their nation and believed in the power of science to transform nature and society into a better world.

National Socialism lacked a consistent economic or social theory. Instead it put forward a series of disjointed policies, subsidies, retrenchments, plans, etc., an agglomeration of ideas in which race, the concept of blood and soil, *Lebensraum*, the leader principle, and so on were prominent in justification. Soviet Marxism developed its economic policies around the proletarian state, service to the nation, and developing industrial might. But both revered technology: for example, the autobahn and the metro, the V-2 rocket and the atomic bomb. And there was little the public or scientists could do to divert the state from its activities as the prime mover behind large-scale technological systems.

Note

1. Many of the ideas in this section are developed in Paul Josephson, " 'Projects of the Century' in Soviet History: Large-Scale Technologies from Lenin to Gorbachev," *Technology and Culture* 36, no. 3 (July 1995): 519–59.

Conclusion: Totalitarian Science?

I N 1949, THE government of dictator Juan Perón announced that Argentinean scientists had succeeded in building a thermonuclear fusion reactor. Scientists in the rest of the world greeted this announcement with justified skepticism, but it was intended for political more than scientific ends. Like many governments, so Perón's recognized the significance of achievements in big science and technology in securing the ideological, cultural, and political legitimacy of the regime.

The reader will have recognized that science and technology have common features under all political systems. From the frequent references in this book to American science, it should be clear that these features include the fact that scientists everywhere strive to establish universal regularities, laws, and theories. They marshal "facts" using hypotheses, deductive reasoning, experiments, and the construction of proofs. Because their research is often expensive, they rely on external public and private sources of funding, including government and industry. Finally, scientists everywhere face obstacles to the smooth conduct of research, from a variety of origins that are both internal to their disciplines and external: moral, ethical, ideological, political, financial, and professional factors.

With explicit, but more often implicit, commentary, I have held up the U.S. scientific enterprise as the paradigmatic liberal pluralist regime—in opposition to totalitarian science. Surely, in the United States there is a national style of science and technology that is influenced by ideology, by politics, and by culture. It is difficult to imagine the honest scientist toiling independently in pursuit of the "absolute truth." J. Robert Oppenheimer's treatment at the hands of the military, FBI, and McCarthyites; the debates over using hormones to increase milk production and over the harmful effects of power lines or silicon implants; the efforts of many in the U.S. Congress to ignore scientific consensus about the nature of wetlands or forest to permit irreversible development: these and many, many other examples demonstrate that "science" (in the sense of what people believe about nature) is socially constructed. Yet public access to scientific and technological decisions (for example, through the environmental and right-to-life movements and through the lively debate over new ideas that takes place in scientific journals and even on television and in the popular press) and the absence of one predominating ideological position distinguish this science from totalitarian science.

If national culture and polity shape the experience of scientists and engineers everywhere, in totalitarian regimes they have the strongest impact. One way in which this occurs is the adoption of a transformationist vision to orient research, a vision that parallels and augments national political aspirations. Whether Lysenkoist (and Lamarckian) reworking of nature, including humans, for a proletarian state, or a determinist longing for a racially pure and thereby thousand-year Reich, the transformationist essence has had a significant impact not only on science and the professional lives of scientists but on society at large, as can be seen in its effects on both millions of unfortunate human victims and the natural environment.

Ideologization also distinguishes science in totalitarian regimes. This may be more likely to occur when advances in science provoke an epistemological crisis that seems somehow to threaten the foundations of state ideology. Such a crisis, for example, accompanied the genesis of the new physics: Marxist philosphers strongly felt the danger from "idealism" as hostile to the working class; Aryan scientists were certain that relativity theory was Jewish physics and as such had no place in German universities or culture. No dissent was tolerated; no public airing of disputes was allowed; and professional organizations were required to show publicly their approval of official ideological positions. Once again, there were many victims, and the progress of science lagged substantially.

Finally, by definition, totalitarian science is "big" science and technology. Whether fascist or socialist economic relations exist, the state is the prime mover behind research, development, and diffusion. Its policy makers and economic planners insist upon a close tie between science and production. They create top-heavy bureaucracies to supervise scientists' activities. In many cases, single individuals and institutions end up dominating entire fields of research. Not only in biology, physics, and chemistry but also in the technological artifacts of society, the result is gigantomania. Under the Nazi and Soviet regimes, there were many examples of successful state planning and execution of technological developments. But there were also great human costs.

Huge monuments whose scale was intended to dwarf human sensibilities were built. Their social and environmental consequences will be with us for decades to come.

In the end, the history of science and technology in totalitarian regimes remains a failed experiment. There were significant social and human costs in the two regimes upon which we have focused, Nazi Germany and the USSR, neither of which survived the century. Science in existing totalitarian regimes merely struggles along. In China, the political authorities have embraced economic reforms to mimic a market economy, while maintaining state prerogatives. They encourage scientists to embark on cutting-edge research in molecular biology and nuclear physics but insist upon crude ideological controls and underfund the R and D effort when they do not perceive short-term economic benefits. Autarky, applied science, and accountability prevail.

The few other totalitarian regimes we might consider—North Korea and Cuba, for example—follow the standard outline I have presented to characterize totalitarian science. They are anomalies, divorced from the international scientific community. Scientists lack autonomy to fix their research programs, since these are tied firmly to state economic goals and ideological precepts. And by such quantitative measures as scientific citations and international awards, their achievements lag. But the goal of this book has not been to denigrate these achievements, nor to hold up the liberal pluralist model as a panacea. Rather it has been to encourage the reader to recognize the potential of political, cultural, and economic institutions for shaping the practice of science.

Bibliography

Adam, Peter. 1994. *Art of the Third Reich.* New York: Harry N. Abrams.

Adams, Mark B. 1968. "The Founding of Population Genetics: Contributions of the Chetverikov School, 1924–1934." *Journal of the History of Biology* 1, no. 1: 23–39.

Adams, Mark B. 1970. "Towards a Synthesis: Population Concepts in Russian Evolutionary Thought, 1925–1935." *Journal of the History of Biology* 3, no. 1: 107–29.

Adams, Mark B. 1980. "Science, Ideology, and Structure: The Kol'tsov Institute, 1900–1970." In *The Social Context of Soviet Science,* ed. Linda Lubrano and Susan Solomon. Boulder: Westview Press.

Advisory Committee on Human Radiation Experiments. 1994. *Interim Report of the Advisory Committee on Human Radiation Experiments.* Washington.

Aitken, Hugh G. J. 1985. *Scientific Management in Action: Taylorism at Watertown Arsenal, 1908–1915.* Princeton: Princeton University Press.

Alexander, Leo. 1949. "Medical Science under Dictatorship." *New England Journal of Medicine* 241, no. 2 (July): 39–47.

Arkhiv Akademii Nauk SSSR, f. 596, op. 2, no. 174.

Arkhiv Akademii Nauk SSSR, f. 351, op. 1, ed. khr. 82.

Arkhiv Akademii Nauk SSSR, f. 351, op. 2, ed. khr. 26.

Arkhiv Lenigradskogo fiziko-tekhnicheskogo instituta, f. 3, op. 1, ed. khr. 195.

Arkhiv Moskovskogo gosudarstvennogo universiteta, f. 1, op. 201.

Bacon, Francis. 1942. *Essays and New Atlantis,* ed. Gordon S. Haight. Toronto, New York, and London: D. Van Nostrand.

Badash, Lawrence. 1985. *Kapitza, Rutherford, and the Kremlin.* New Haven: Yale University Press.

Badash, Lawrence. 1995. *Scientists and the Development of Nuclear Weapons: From Fission to the Limited Test Ban Treaty, 1939–1963.* Atlantic Highlands, NJ: Humanities Press.

Bailes, Kendall. 1977. "Alexei Gastev and the Soviet Controversy over Taylorism, 1918–1924." *Soviet Studies* 29, no. 3 (July): 373–94.

Bailes, Kendall. 1978. *Technology and Society under Lenin and Stalin.* Princeton: Princeton University Press.

Bailes, Kendall. 1990. *Science and Russian Culture in an Age of Revolutions.* Bloomington: Indiana University Press.

Banham, Reyner. 1986. *Theory and Design in the First Machine Age.* 2d ed. Cambridge: MIT Press.

Berger, Robert. 1990. "Nazi Science—The Dachau Hypothermia Experiments." *New England Journal of Medicine* 322, no. 20: 1435–40.

Bernstein, Jeremy. 1993. "Revelations from Farm Hall." *Science* 259 (26 March): 1993–96.

111

Beyerchen, Alan D. 1977. *Scientists under Hitler: Politics and the Physics Community in the Third Reich*. New Haven and London: Yale University Press.

Boag, J. W., P. E. Rubinin, and D. Shoenberg. 1990. *Kapitza in Cambridge and Moscow: Life and Letters of a Russian Physicist*. Amsterdam, Oxford, and New York: North-Holland.

Bracher, Karl Dietrich. 1970. *The German Dictatorship*, tr. Jean Steinberg. New York and Washington: Praeger Publishers.

Bressler, Michael L. 1992. *Agenda Setting and the Development of Soviet Water Resources Policy, 1965–1990: Structures and Processes*. Ph.D. dissertation, University of Michigan, Ann Arbor.

Burleigh, Michael. 1995. *Death and Deliverance: "Euthanasia" in Germany c. 1900–1945* Cambridge: Cambridge University Press.

Cassidy, David. 1992. *Uncertainty: The Life and Science of Werner Heisenberg*. New York: W. H. Freeman.

Cassidy, David. 1995. *Einstein and Our World*. Atlantic Highlands, NJ: Humanities Press International.

Conquest, Robert. 1968. *The Great Terror*. New York: Collier Books.

Coopersmith, Jonathan. 1992. *The Electrification of Russia, 1880–1926* Ithaca: Cornell University Press.

Dalrymple, Dana. 1963. "The American Tractor Comes to Soviet Agriculture: The Transfer of a Technology." *Technology and Culture* 5: 191–214.

Darré, R. Walther. 1933. *Das Schwein als Kriterium für nordische Völker und Semiten*. Munich: J. Lehmanns Verlag.

Darré, R. Walther. 1934. *Im Kampf um die Seele des deutschen Bauern*. Berlin: P. Steegemann.

Darré, R. Walther. 1938. *Das Bauerntum als Lebensquell der Nordischen Rasse*. Munich and Berlin: J. Lehmanns Verlag. Orig. pub. 1928.

Deichmann, Ute, and Benno Muller-Hill. 1994. "Biological Research at Universities and Kaiser Wilhelm Institutes in Nazi Germany." In *Science, Technology, and National Socialism*, ed. Monika Renneberg and Mark Walker. Cambridge: Cambridge University Press.

Dobbs, Betty Jo Teeter, and Margaret C. Jacob. 1995. *Newton and the Culture of Newtonianism*. Atlantic Highlands, NJ: Humanities Press International.

Ellul, Jacques. 1964. *The Technological Society*. New York: Vantage.

Enkena, V. 1962. "Pamiati uchitel'ia." *Za nauku v sibiri* 51 (76) (December 12): 4.

Fang, Lizhi. 1990. *Bringing Down the Great Wall*. New York and London: Norton.

Feshbach, Murray, and Alfred Friendly Jr. 1992. *Ecocide in the USSR*. New York: Basic Books.

Fishman, Robert. 1982. *Urban Utopias in the Twentieth Century*. Cambridge: MIT Press.

Folop-Mueller, Rene. 1965. *The Mind and Face of Bolshevism*. Ann Arbor: University Microfilms.

Forman, Paul. 1971. "Weimar Culture, Causality, and Quantum Theory, 1918–1922." *Historical Studies in the Physical Sciences* 3: 1–115.

Forman, Paul. 1974. "The Financial Support and Political Alignment of Physicists in Weimar Germany." *Minerva* 12: 39–66.

Frederiks V. K., and A. A. Fridman. 1924. *Osnovy teorii otnositel'nosti*. Leningrad: n.p.

Frenkel', Ia. I. 1923. *Teoriia otnositel'nosti* Petrograd: n.p.

Frost, Robert. 1991. *Alternating Currents: Nationalized Power in France, 1946–1971.* Ithaca: Cornell University Press.

Gamow, George. 1970. *My World Line.* New York: Viking Press.

Gay, Peter. 1968. *Weimar Culture.* New York: Harper.

Gessen, Boris, 1930. "K voprosu o probleme prichinnosti v kvantovoi mekhanike." In *Volny materii i kvantovaia mekhanika,* ed. Artur Gass, tr. P. S. Tartakovskii. Moscow and Leningrad: n.p.

Gessen, Boris. 1971. "The Socio-economic Roots of Newton's Principia." In *Science at the Crossroads,* ed. Herbert Dingle. London: Martin, Brian and O'Keefe Ltd.

Giles, Geoffrey. 1992. "'The Most Unkindest Cut of All': Castration, Homosexuality, and Nazi Justice." *Journal of Contemporary History* 27: 41–61.

Gimbel, John. 1990. "German Scientists, United States Denazification Policy, and the '*Paperclip* Conspiracy.'" *International History Review* 12, no. 3 (August): 441–65.

Gleason, Abbott. 1995. *Totalitarianism: The Inner History of the Cold War.* New York and Oxford: Oxford University Press.

Goebbels, Joseph. 1948. *The Goebbels Diaries,* tr. and ed. Louis Lochner. London: Hamish Hamilton.

Goldman, Marshall. 1972. *The Spoils of Progress: Environmental Pollution in the Soviet Union.* Cambridge: MIT Press.

Gorelik, Gennady, and Victor Frenkel'. 1994. *Matvei Petrovich Bronstein and Soviet Theoretical Physics in the Thirties,* tr. Valentina Levina. Basel, Boston, and Berlin: Birkhäuser Verlag.

Göring, Hermann. 1939. *Political Testament of Hermann Göring,* tr. H. W. Blood-Ryan. London: John Long.

Gorky, M., L. Averbakh, and S. Firin. 1934. *Belomorsko-baltiiskii kanal imeni Stalina. Istoriia stroitel'stva.* Moscow: Gosizdat 'istoriia fabrik i zavodov.'

Goudsmit, Samuel A. 1947. *ALSOS* New York: Henry Schuman.

Gould, Stephen Jay. 1981. *The Mismeasure of Man.* New York: Norton.

Graham, Loren. 1967. *The Soviet Academy of Sciences and the Communist Party, 1927–1932.* Princeton: Princeton University Press.

Graham, Loren. 1981. *Between Science and Values.* New York: Columbia University Press.

Graham, Loren. 1985. "The Socio-political Roots of Boris Hessen: Soviet Marxism and the History of Science." *Social Studies of Science* 15: 705–22.

Graham, Loren. 1987. *Science, Philosophy, and Human Behavior in the Soviet Union.* New York: Columbia University Press.

Graham, Loren. 1992. "The Fits and Starts of Russian and Soviet Technology. In *Technology, Culture, and Development: The Experience of the Soviet Model,* ed. James Scanlan. Armonk: M. E. Sharpe.

Graham, Loren. 1993. *The Ghost of the Executed Engineer.* Cambridge: Harvard University Press.

Gropius, Walter. 1965. *The New Architecture and the Bauhaus.* Cambridge: MIT Press.

Gustafson, Thane. 1981. *Reform in Soviet Politics: Lessons of Recent Policies on Land and Water.* Cambridge: Cambridge University Press.

Heilbron, John. 1986. *The Dilemmas of an Upright Man.* Berkeley, Los Angeles, and London: University of California Press.

Heinemann-Grüder, Andreas. 1994. "Keinerlei Untergang: German Armaments

Engineers during the Second World War and in Service of the Victorious Powers." In *Science, Technology, and National Socialism*, ed. Monika Renneberg and Mark Walker. Cambridge: Cambridge University Press.

Herf, Jeffrey. 1984. *Reactionary Modernism: Technology, Culture, and Politics in Weimar and the Third Reich.* Cambridge and New York: Cambridge University Press.

Hiebert, Erwin. 1979. "The State of Physics at the Turn of the Century." In *Rutherford and Physics at the Turn of the Century*, ed. Mario Bunge and William Shea. Kent, England: Dawson; New York: Science History Publications.

Holloway, David. 1994. *Stalin and the Bomb: The Soviet Union and Atomic Energy, 1939–1956.* New Haven and London: Yale University Press.

Holt, John. 1936. *German Agricultural Policy, 1918–1934.* New York: Russell and Russell.

Hughes, Thomas P. 1969. "Technological Momentum in History: Hydrogenation in Germany, 1898–1933." *Past and Present* 44 (August): 106–32.

Hughes, Thomas P. 1983. *Networks of Power.* Baltimore: Johns Hopkins University Press.

Hughes, Thomas P. 1989a. "The Evolution of Large Technological Systems." In *The Social Construction of Technological Systems*, ed. Wiebe E. Bijker, Thomas P. Hughes, and Trevor Pinch. Cambridge: MIT Press.

Hughes, Thomas P. 1989b. *American Genesis.* New York: Viking.

Ioffe, Abram, ed. 1927. *Osnovaniia novoi kvantovoi mekhaniki.* Moscow and Leningrad: n.p.

Ioffe, Abram. 1930. "Korennye problemy NIR." In *Sotsialisticheskaia rekonstruktsiia i nauchno-issledovatel'skaia rabota. Sbornik NIS PTEU VSNKh SSSR k XVI s"ezdu VKP (b).* Moscow: n.p.

Ioffe, Abram. 1937. "O polozhenii na filosofskom fronte sovetskoi fiziki." *Pod znamenem marksizma* 11–12: 133–43.

Izvestiia Akademii Nauk SSSR. Seriia fizicheskaia 1936. 1–2.

Jones, James. 1993. *Bad Blood.* Rev. ed. New York: Free Press.

Joravsky, David. 1961. *Soviet Marxism and Natural Science, 1917–1931.* New York: Columbia University Press.

Josephson, Paul. 1981. "Science and Ideology in the Soviet Union: The Transformation of Science into a Direct Productive Force." *Soviet Union.* 8 (pt. 2): 159–185.

Josephson, Paul. 1986. "The Historical Roots of the Chernobyl Crisis." *Soviet Union* 13, no. 3: 275–99.

Josephson, Paul. 1987. "The Early Years of Soviet Nuclear Physics." *Bulletin of the Atomic Scientists* 43, no. 10 (December): 36–39.

Josephson, Paul. 1991. *Physics and Politics in Revolutionary Russia.* Los Angeles and Oxford: University of California Press.

Josephson, Paul. 1995. "'Projects of the Century' in Soviet History: Large-Scale Technologies from Lenin to Gorbachev." *Technology and Culture* 36, no. 3 (July): 519–59.

Kaganovich, Lazar. 1931. *Za sotsialisticheskuiu rekonstruktsiiu Moskvy i gorodov SSSR.* Moscow and Leningrad: OGIZ 'Moskovskii rabochii.'

Kaganovich, L. M. 1935. *Pobeda metropolitena—pobeda sotsializma.* Moscow: Transzheldorizdat.

Kaganovich, Lazar. 1936. *Voprosy zheleznodorozhnogo transporta v sviazi so stakhanovskim dvizheniem.* Moscow: Transzheldorizdat.

Kaptisa, P. L. 1989. *Pis'ma o nauke, 1930–1980.* Moscow: Moskovskii rabochii.
Kataev, Valentin. 1976. *Time, Forward,* tr. Charles Malamuth. Bloomington and London: Indiana University Press.
Kattsen, I. 1947. *Metro Moskvy.* Moscow: Moskovskii rabochii.
Kennedy, Foster. 1942. "The Problem of Social Control of the Congenital Defective." *American Journal of Psychiatry* 99: 13–16.
Kevles, Daniel. 1979. *The Physicists.* New York: Vintage Books.
Kevles, Daniel. 1985. *In the Name of Eugenics.* Berkeley and Los Angeles: University of California Press.
Kohler, Robert. 1991. *Partners in Science: Foundations and Natural Scientists, 1900–1945.* Chicago: University of Chicago Press.
Kopp, Anatole. 1970. *Town and Revolution: Soviet Architecture and City Planning, 1917–1935,* tr. Thomas Burton. New York: G. Braziller.
Koritskii, E. B. ed. 1990. *U istokov NOT. Zabytye diskussii i nerealizovannye idei.* Leningrad: Leningrad University.
Kosarev, V. V. 1993. "Fiztekh, gulag i obratno." In *Chteniia pamiati A. F. Ioffe, 1990,* ed. V. M. Tuchkevich. St. Petersburg: Nauka.
Koshelev, F. P. 1952. *Velichestvennye stalinskie stroiki kommunizma i ikh narodnokhoziaistvennoe znachenie.* Moscow: Gosizdatpolit.
Kotkin, Stephen. 1991. *Steeltown, USSR.* Berkeley and Los Angeles: University of California Press.
Kravetz, T. P. 1928. "Shestoi s"ezd fizikov." *Priroda* 10: 915–16.
Kuleshov, N., and A. Pozdnev. 1954. *Vysotnye zdaniia Moskvy.* Moscow: Moskovskii rabochii.
Kuromiya, Hiroaki. 1988. *Stalin's Industrial Revolution: Politics and Workers, 1928–1932.* Cambridge and New York: Cambridge University Press.
Lane, Barbara Miller. 1985. *Architecture and Politics in Germany, 1918–1945.* Cambridge and London: Harvard University Press.
Lenard, Philipp. 1920. *Über Relativitätsprinzip, Äther, Gravitation* Leipzig: Verlag S. Hirzel.
Lenard, Philipp. 1933. *Great Men of Science: A History of Scientific Progress,* tr. H. Stafford Hatfield. New York: Macmillan.
Lenin, V. I. 1956. *The Development of Capitalism in Russia.* Moscow: Foreign Language Publishing House.
Lennox, William. 1938. "Should They Live?" *American Scholar* 7: 454–66.
Levshin, V. L. 1936. "Novye puti sovetskoi fiziki." *Vestnik Akademii Nauk SSSR* 4–5: 62–75.
Lifton, Robert J. 1986. *The Nazi Doctors.* New York: Basic Books.
Lissitzky, El. 1984. *Russia: An Architecture for World Revolution,* tr. E. Dluhosch. Cambridge: MIT Press.
Macrakis, Kristie. 1994. *Surviving the Swastika.* New York and Oxford: Oxford University Press.
Makovskii, V. L. 1945. "Moskovskii metropoliten." *Nauka i zhizn'* 7: 40–45.
Marchuk, Gurii. 1991. "Vystuplenie na pervoi konferentsii unchenykh nauchnykh uchrezhdenii Rossiiskoi Akademii Nauk." 10 December. Typescript.
Mazumdar, Pauline M. H. 1990. "Blood and Soil: The Serology of the Aryan Racial State." *Bulletin of the History of Medicine* 64: 187–219.
Mehrtens, Herbert. 1994. "Irresponsible Purity: The Political and Moral Structure of the Mathematical Sciences in the National Socialist State." In *Science, Technology, and National Socialism,* ed. Monika Renneberg and Mark Walker.

Cambridge: Cambridge University Press.

Mekhanisticheskoe estestvoznania i dialekticheskii materializm. 1925. Vologda: n.p.

Merton, Robert. 1973. "The Normative Structure of Science." In Robert Merton, *The Sociology of Science.* Chicago and London: University of Chicago Press.

Miller, Robert. 1970. *100,000 Tractors: The MTS and the Development of Controls in Soviet Agriculture.* Cambridge: Harvard University Press.

Moiseenko, V. P., et al., eds. 1990. *Razvitie stroitel'noi nauki i tekhniki v ukrainskoi SSSR,* vols. 2 and 3. Kiev: Naukova dumka.

Murray, Charles, and Richard Herrnstein. 1994. *The Bell Curve.* New York: Free Press.

Nauchno-organizatsionnaia deiatel'nosti Akademika A. F. Ioffe. Sbornik dokumentov. 1980. Leningrad: Nauka.

Neufeld, Michael. 1990. "Weimar Culture and Futuristic Technology: The Rocketry and Spaceflight Fad in Germany, 1923–33." *Technology and Culture* 31: 725–52.

Neufeld, Michael J. 1994. "The Guided Missile and the Third Reich: Peenemünde and the Forging of a Technological Revolution." In *Science, Technology, and National Socialism,* ed. Monika Renneberg and Mark Walker. Cambridge: Cambridge University Press.

Neumann, Franz. 1942. *Behemoth.* New York: Harper and Row.

Nicholson, Heather Johnston. 1977. "Autonomy and Accountability of Basic Research." *Minerva* 15, no. 1 (Spring): 32–61.

Noakes, Jeremy. 1984. "Nazism and Eugenics: The Background to the Nazi Sterilization Law of 14 July 1933." In *Ideas into Politics: Aspects of European History, 1880 to 1950,* ed. R. J. Bullen, H. Pogge von Strandmann, and A. B. Polonsky. Totowa: Barnes and Nobles.

Orlov, Yuri. 1991. *Dangerous Thoughts,* tr. Thomas P. Whitney. New York: William Morrow.

"Ot redaktsii." *Dialektika v prirode,* sb. 2, p. i.

Overy, R. J. 1975. "Cars, Roads, and Economic Recovery in Germany, 1932–8." *Economic History Review* 28: 466–83.

Pfetsch, Frank. 1970. "Scientific Organization and Science Policy in Imperial Germany, 1871–1914: The Foundation of the Imperial Institute of Physics and Technology." *Minerva* 8 (October): 557–80.

Pois, Robert. 1970. "Introduction." In Alfred Rosenberg, *Selected Writings,* ed. Robert Pois. London: Jonathan Cape.

Polanyi, Michael. 1962. "The Republic of Science." *Minerva* 1: 54–73.

Popper, Karl. 1942. *The Open Society and Its Enemies.* London: G. Routledge and Son.

Powers, Thomas. 1993. *Heisenberg's War: The Secret History of the German Bomb.* New York: Knopf.

Proctor, Robert. 1988. *Racial Hygiene: Medicine under the Nazis.* Cambridge and London: Harvard University Press.

Rassweiler, Anne D. 1988. *The Generation of Power.* New York and Oxford: Oxford University Press.

Renneberg, Monika, and Mark Walker. 1994. "Scientists, Engineers, and National Socialism." In *Science, Technology and National Socialism,* ed. Monika Renneberg and Mark Walker. Cambridge: Cambridge University Press.

Rhodes, Richard. 1995. *Dark Sun: The Making of the Hydrogen Bomb.* New York and London: Simon and Schuster.

Rosenberg, Alfred. 1970. *Selected Writings*, ed. and introduced by Robert Pois. London: Jonathan Cape.

Rössler, Mechtild. 1994. "'Area Research' and 'Spatial Planning' from the Weimar Republic to the German Federal Republic: Creating a Society with a Spatial Order under National Socialism." In *Science, Technology, and National Socialism*, ed. Monika Renneberg and Mark Walter. Cambridge: Cambridge University Press.

Ruble, Blair. 1994. "Failures of Centralized Metropolitanism: Inter-war Moscow and New York." *Planning Perspectives* 9: 353–76.

Sakharov, Andrei. 1990. *Memoirs*, tr. Richard Lourie. New York: Knopf.

Schorske, Carl. 1979. *Fin-de-siècle Vienna: Politics and Culture.* New York: Knopf.

Scott, John. 1989. *Behind the Urals: An American Worker in Russia's City of Steel.* Bloomington and Indianapolis: Indiana University Press. Orig. pub. 1942.

Segrè, Emilio. 1980. *From X-rays to Quarks: Modern Physicists and Their Discoveries.* San Francisco: W. H. Freeman.

Shallat, Todd. 1994. *Structures in the Stream: Water, Science, and the Rise of the U.S. Army Corps of Engineers.* Austin: University of Texas Press.

Sherwin, Martin. 1975. *A World Destroyed.* New York: Alfred A. Knopf.

Siegelbaum, Lewis. 1984. "Soviet Norm Determination in Theory and Practice, 1917–1941." *Soviet Studies* 36 (January): 45–68.

Siegelbaum, Lewis. 1988. *Stakhanovism and the Politics of Productivity in the USSR, 1935–1941.* Cambridge: Cambridge, University Press.

Smil, Vaclav. 1993. *China's Environmental Crisis.* Armonk, NY: M. E. Sharpe.

Smith, Michael L. 1983. "Selling the Atom: The U.S. Manned Space Program and the Triumph of Commodity Scientism." In *The Culture of Consumption: Critical Essays in American History, 1880–1980,* ed. Richard Fox and T. J. Jackson Lears. New York: Pantheon.

Smyth, Henry. 1946. *Atomic Energy for Military Purposes.* Princeton: Princeton University Press.

Speer, Albert. 1970. *Inside the Third Reich,* tr. Richard and Clara Winston. New York: Collier Books.

Stark, Johannes. 1932. *Adolf Hitlers Ziele und Persönlichkeit.* Munich: Deutscher Volksverlag.

Stark, Johannes. 1934. *Nationalsozialismus und Wissenschaft.* Munich: Zentralverlag der NSDAP.

Staudenmaier, J. M. 1974. "An Electrifying Leader: Lenin's Leadership in Technology and Electrification, 1920–1923." Unpublished paper, University of Pennsylvania, Philadelphia.

Stites, Richard. 1989. *Revolutionary Dreams.* New York and Oxford: Oxford University Press.

Taubes, Gary. 1986. *Nobel Dreams.* New York: Random House.

Teich, Albert, and W. Henry Lambright. 1976–77. "The Redirection of a Large National Laboratory." *Minerva* 14, no. 4 (Winter): 447–74.

Todt, Fritz. 1932. *Strassenbau und Strassenverwaltung.* Munich: n.p.

Tushkin, A. A., ed. 1988. *Teoriia i metody upravleniia vodnymi resursami sushi,* part 1. Moscow: ONK VASKhNIL.

Vecoli, Rudolph. 1960. "Sterilization: A Progressive Measure?" *Wisconsin Magazine of History* 43: 190–202.

"Velikie stroiki kommunizma." 1951. *Elektrosila* 9. Leningrad and Moscow: Gosenergoizdat.

Velikie stroiki stalinskoi epokhi. Kratkii rekomendatel'nyi ukazatel' literatury. 1950. Moscow: Biblioteka im. Lenina.

Verluise, Pierre. 1995. *Armenia in Crisis: The 1988 Earthquake,* tr. Levon Chorbajian. Detroit: Wayne State University Press.

Vinter, A. V. 1951. *Velikie stroiki kommunizma.* Moscow: Izdatel'stvo Akademii Nauk SSSR.

Voropaev, G. V., and D. Ia. Ratkovich. 1985. *Problema territorial'nogo pereraspredeleniia vodnykh resursov.* Moscow: IVP AN SSSR.

Walker, Mark. 1989a. "National Socialism and German Physics." *Journal of Contemporary History* 24:

Walker, Mark. 1989b. *National Socialism and the Quest for Nuclear Power.* Cambridge: Cambridge University Press.

Weiner, Douglas. 1988. *Models of Nature: Ecology, Conservation, and Cultural Revolution in Soviet Russia.* Bloomington: Indiana University Press.

Weiner, Douglas. 1992. "Demythologizing Environmentalism." *Journal of the History of Biology* 25, no. 3 (Fall): 385–412.

Weiss, Sheila Faith. 1994. "Pedagogy, Professionalism, and Politics: Biology Instruction during the Third Reich." In *Science, Technology, and National Socialism,* ed. Monika Renneberg and Mark Walker. Cambridge: Cambridge University Press.

Wells, H. G. 1921. *Russia in the Shadows.* New York: George H. Doran.

Whitford, Frank. 1984. *Bauhaus.* London: Thames and Hudson.

Zamiatin, Eugene. 1924. *We,* tr. Gregory Zilboorg. New York: E. P. Dutton.

Zhukovskii, P. M., D. K. Beliaev, and S. I. Alikhanian. 1972. "50 let otechestvennoi genetiki i selektsii rastenii, zhivotnykh i mikroorganizmov." *Genetika* 8, no. 12: 21.

Zhurnal tekhnicheskoi fiziki. 1973. 8: 884.

Index

119